Energy Systems in Electrical Engineering

Series editor

Muhammad H. Rashid, Pensacola, USA

More information about this series at http://www.springer.com/series/13509

Anindya Dasgupta · Parthasarathi Sensarma

Design and Control of Matrix Converters

Regulated 3-Phase Power Supply and Voltage Sag Mitigation for Linear Loads

Springer

Anindya Dasgupta
Avionics Department
Indian Institute of Space Science and Technology
Thiruvananthapuram, Kerala
India

Parthasarathi Sensarma
Department of Electrical Engineering
Indian Institute of Technology Kanpur
Kanpur, Uttar Pradesh
India

ISSN 2199-8582 ISSN 2199-8590 (electronic)
Energy Systems in Electrical Engineering
ISBN 978-981-10-3829-7 ISBN 978-981-10-3831-0 (eBook)
DOI 10.1007/978-981-10-3831-0

Library of Congress Control Number: 2017931048

Printed on acid-free paper

This Springer imprint is published by Springer Nature
The registered company is Springer Nature Singapore Pte Ltd.
The registered company address is: 152 Beach Road, #21-01/04 Gateway East, Singapore 189721, Singapore

Preface

Improving power density of converters, apart from conversion efficiency, has been one of the major drivers of modern developments in power electronics. Tracing the evolution of power electronics reveals that improvement in semiconductor devices, digital control with ever-increasing computational capability, improved packaging techniques combined with parallel development of converter topologies, switching schemes and advanced control techniques have kept on redefining the benchmarks with every passing day. The power converter which forms the core of any power electronic application is an electrical network involving only semiconductor switches and passive storage elements, with the latter largely contributing to the volume and weight of the converter. Therefore it is quite natural that, among competing candidates for a given application, the topology with the highest semiconductor to passives ratio—both in terms of component count and size—would offer the highest power density. For this specific possibility, the Matrix Converter—a generic name for any converter having only semiconductor devices in its power stage—demands focused investigation.

Matrix converters were proposed in the 1980s and almost immediately generated a lot of research interest, mostly in the domain of 3-phase AC-AC power conversion, and there has been several significant breakthroughs in modulation methods, multi-step commutation and space vector pulse width modulation. With respect to applications, academic and industrial research interest in Matrix Converters has been largely confined to motor drives. But since the last decade, there had been an emerging interest towards using Matrix Converters in power system applications. Most of these applications demand faster dynamic performance than the industrial drives, which have high plant inertia and thus do not require such response speeds. Absence of any intermediate energy buffering element makes the Matrix Converter a dynamically tightly coupled input–output unit and the overall system design, particularly controller design, challenging.

The research efforts which led to this book were a part of this effort to expand the application domain of Matrix Converter to power systems. Two target applications for synchronous systems have been addressed—regulated 3-Phase voltage supply and voltage sag mitigation. The objective of the book has been subsequently categorized into the following—developing a dynamic model which provides adequate design insight, filter design and devising a control scheme. The low-frequency dynamic model is first analysed for regulated voltage supply assuming balanced system. A linearized dynamic model is developed and it is shown that depending on the input power, input voltage and filter parameters a possibility of appearance of a set of right half zeros exist.

The design of filters is considered next. Apart from general issues like ripple attenuation, regulation, reactive current loading and filter losses, additional constraints which may be imposed by dynamic requirements and commutation are also addressed.

In the third stage, voltage controller design is detailed for 3-Phase regulated voltage supply. In the synchronous *dq* domain, output voltage control represents a multivariable control problem. The control problem is reduced to a single variable one while retaining all possible right half zeros, thereby preserving the internal stability of the system. Consequently, standard single variable control design technique has been used to design a controller. The analytically predicted dynamic response has been verified by experimental results. The system could be operated beyond the critical power boundary where the right half zeros emerge.

Finally, the developed control approach has been extended to voltage sag mitigation with adequate modifications. A 3-wire linear load has been considered. Both symmetrical and asymmetrical voltage sags have been considered.

Thiruvananthapuram, India Anindya Dasgupta
Kanpur, India Parthasarathi Sensarma

Contents

About the Authors

Dr. Anindya Dasgupta is an assistant professor of Avionics at the Indian Institute of Space Science and Technology, Thiruvananthapuram. Dr. Dasgupta received his Ph.D. (Electrical Engineering) from the Indian Institute of Technology (IIT) Kanpur, India. He completed his M.E. (Control System, Electrical Engineering) in 2006 at Bengal Engineering and Science University, Shibpur, and B.E. (Electrical Engineering) in 2000 at the Jadavpur University, Kolkata.

Dr. Parthasarathi Sensarma is an associate professor of Electrical Engineering at IIT Kanpur, India. Dr. Sensarma completed his Ph.D. (Electrical Engineering) in 2001 at the Indian Institute of Science (IISc), Bangalore, India. He was awarded his M.Tech. (Machine Drives and Power Electronics) in 1992 from IIT Kharagpur and B.E.E. (Electrical Engineering) in 1990 from Jadavpur University, Kolkata. He has supervised several Ph.D. and M.Tech. students. Dr. Sensarma has filed patent application at the Indian Patent Office for seven of his projects. He has published 56 technical papers in national and international conferences and journals.

List of Figures

List of Tables

Chapter 1
Introduction

Matrix Converter (MC) is a class of switched mode power converters (SMPC) which contains only semiconductor devices in its power processing stage. These converters were conceived as an alternative to the Back to back inverter (B2BI) topology used in 3-phase to 3-phase power conversion. The necessity of finding an alternative to B2BI in the form of a more silicon reliant solution can be understand by analyzing its power stage. Figure 1.1a, shows the power conversion stage of B2BI.

The power stage of B2BI is made up of two inverters connected back to back through a capacitor which is also commonly referred to as the DC link capacitor. The inverter at the input side transforms the 3-phase sinusoidal AC voltage applied to its input terminals to a switched DC voltage which appears across the DC capacitor. The inverter at the output side requires a very stiff DC link voltage at its input terminals. Consequently a large electrolytic capacitor has to be used as the DC link capacitor. A significant portion of volume and weight of the converter is attributed to the DC link capacitor. Moreover, it has a higher failure rate than the semiconductors. Therefore, a 3-phase AC to AC converter whose power stage is realized only with semiconductor devices is likely to have a higher power density and lifetime than the B2BIs. This was the motivation behind finding a more semiconductor reliant mechanism for power conversion.

Topologically MC is similar to buck converter. Although the former is mostly associated with AC power and the latter strictly with DC, a look at the evolution of a buck converter helps to appreciate how MCs work. Figure 1.2 shows the power stage of a buck converter where the single pole double throw switch Sw is periodically connected to the positive and negative terminals of the input DC voltage source V_{in}. This results in a pulsating voltage v_{sw} at the output. The DC component of v_{sw} can be obtained by averaging the pulse waveform over the switching period T_s as

$$V_{sw} = \frac{1}{T_s}\int_0^{T_s} v_{sw}dt = dV_{in}. \tag{1.1}$$

A. Dasgupta and P. Sensarma, *Design and Control of Matrix Converters*,
Energy Systems in Electrical Engineering, DOI 10.1007/978-981-10-3831-0_1

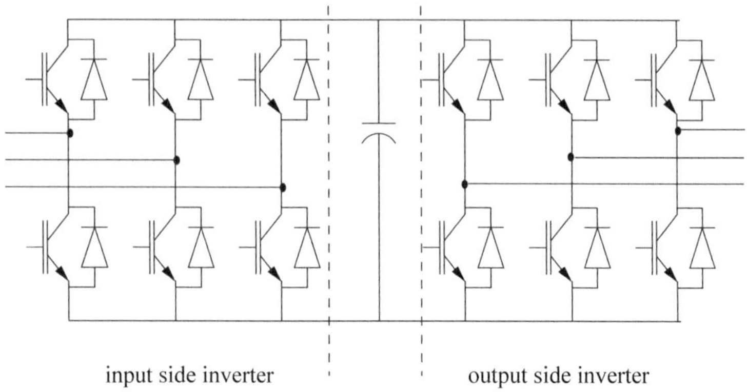

Fig. 1.1 Power processing stage of 3 phase to 3 phase back to back inverters

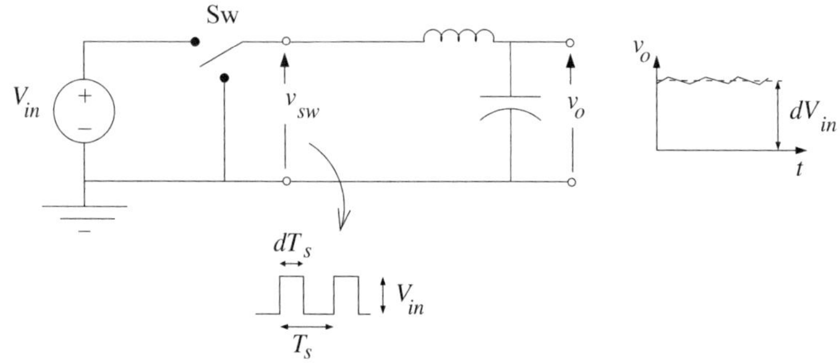

Fig. 1.2 Buck converter realized with a single pole double throw switch

Thus, by regulating the duty cycle d it is possible to synthesize a desirable DC component of the switched voltage v_{sw}. The switching frequency ($f = 1/T_s$) components and its harmonics which are also present in v_{sw} can be subsequently attenuated to a large extent by using a LC filter. The important point to note here is that any desirable value of the DC component of the output voltage can be generated as long as this value is less than the source voltage V_{in}.

Another useful aid in the analysis of 3-phase MC is the output voltage of a 3-phase AC to DC rectifier. Figure 1.3a shows a diode rectifier and the output voltage is shown in Fig. 1.3b. Let the input voltages be

$$v_{sa} = sin2\pi 50t, \quad v_{sb} = sin\,(2\pi 50t - 120°) \quad \& \quad v_{sc} = sin\,(2\pi 50t + 120°)\,. \tag{1.2}$$

Replacing the 3 diodes connected to each of the output terminals by a fully controllable single pole-triple throw(SPTT) switch results in the converter shown in configuration Fig. 1.4. S1 and S2 are the two SPTT switch. Presence of the two SPTT switches makes it possible to synthesize a sinusoidal voltage at the output terminals. The pole of each SPTT switch is connected to the output terminal which ensures that

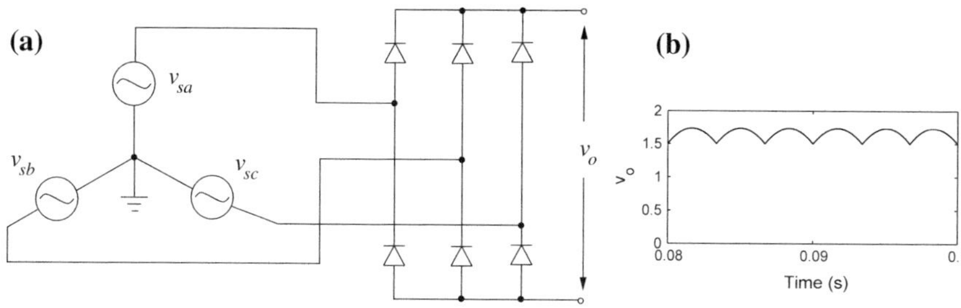

Fig. 1.3 **a** 3-phase diode rectifier. **b** Output voltage

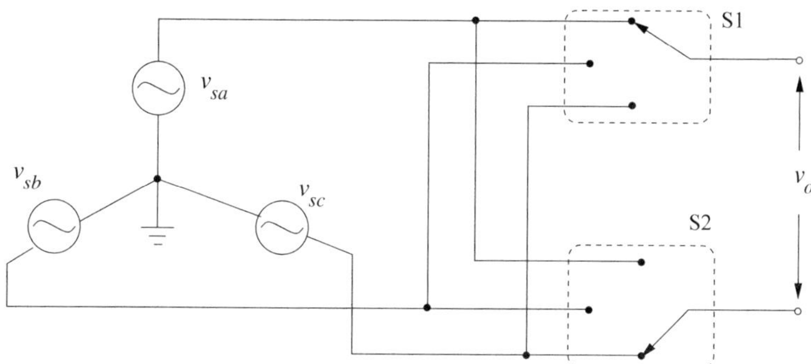

Fig. 1.4 A 3-phase to 1-phase AC converter power stage

the output is always connected to any one of the input voltages. It ensures that the path of the current through the load is never interrupted—a mandatory requirement for inductive load. Figure 1.5 shows the envelope of the input voltages for all possible switching combinations. Taking cue from the operation of buck converter, it should be possible in a similar manner to generate the desired fundamental component of a target AC voltage from the 3-phase source. If the turn-on and turn-off durations of the individual SPTT switches can be properly regulated it is possible to synthesize an output voltage v_o with a desired fundamental frequency component. The only restriction towards achieving this is that the fundamental component must be completely encapsulated within the input voltage envelope. Referring to Fig. 1.5, this restriction implies that the maximum amplitude of v_o is 0.866 times that of the input line-line voltage amplitude. Within this amplitude limit, a fundamental component of v_o having any amplitude, phase and frequency can be produced by appropriately regulating the SPTT switches.

The mode of operating a 3 to 1-phase AC converter can be extended to 3-phase to 3-phase AC conversion as well. Figure 1.6 shows a 3-phase to 3-phase AC to AC converter structure made up of 3 SPTT switch. Since the 3-phase output voltages have to be synthesized directly from the 3 input phases all possible 3-phase output voltage envelopes must be completely encapsulated by the input voltage envelope. Consequently, the maximum achievable gain in voltage amplitude is 0.866. Apart

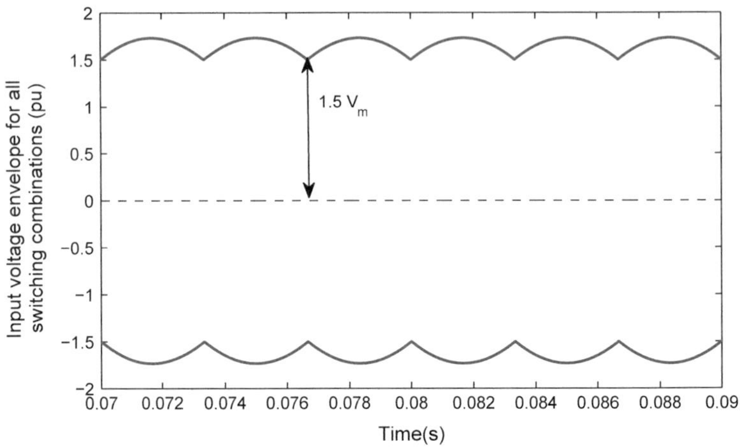

Fig. 1.5 Envelope of the input voltage waveform

Fig. 1.6 Power processing stage of a 3 phase to 3 phase AC-AC converter

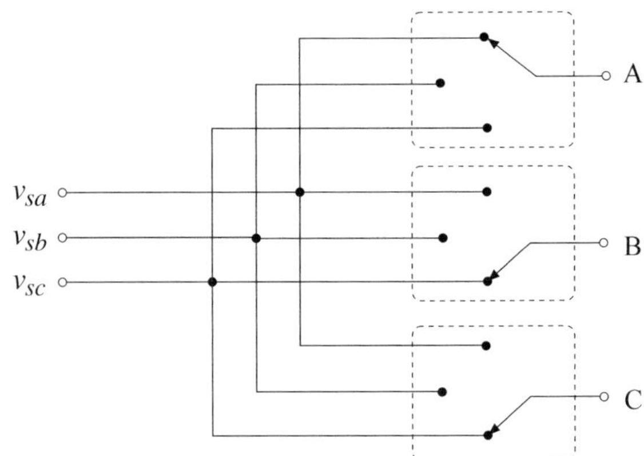

from this fundamental limitation, there is no theoretical limitation on the possible set of output voltage waveforms. Therefore, the operational requirements are simply—realization of the SPTT switches by using available semiconductor switches and devising an appropriate scheme for controlling them.

Each SPTT switch can be realized by 3 single pole-single throw (SPST) switch. By replacing the SPTT switch by 3 SPST switch with common poles connected to an output phase, the topology shown in Fig. 1.7 is obtained. The 9 SPST appears to be arranged in a 3 × 3 matrix and hence the topology is referred to as Matrix converter.

The input terminals of the MC are connected to voltage source and consequently the output terminals has to be connected either resistive or inductive load. The nature of the source and loads which are mostly inductive sets up the following fundamental switching law which must be obeyed

- input terminals should never be shorted,
- output terminals should never be left opened.

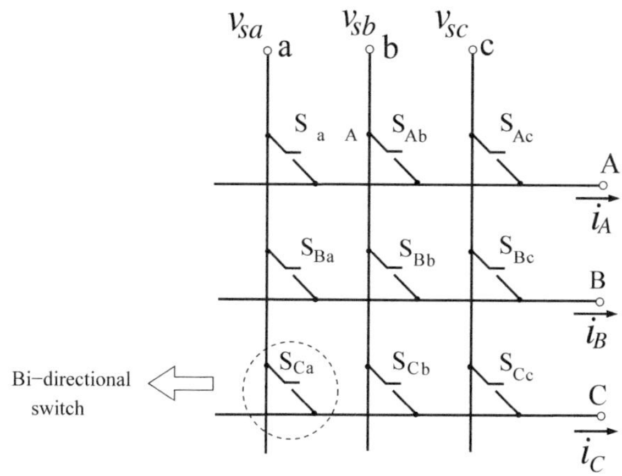

Fig. 1.7 Power processing stage of a 3 phase to 3 phase MC

These constraints can be expressed in terms of switching functions as

$$S_{ja}(t) + S_{jb}(t) + S_{jc}(t) = 1, \quad \text{where,} \quad j \in \{A, B, C\}. \tag{1.3}$$

At any instant, a switching function $S_{xy}(t)$ is either 0 if the switch S_{xy} is open or 1 if it is closed. There can be 27 ($3 \times 3 \times 3$) switching combinations for connecting the output with input side, which obey these fundamental switching rules.

Figure 1.7 represents the conceptual building block of MC topology. Realization of a fully operational converter involves realization of the switches by means of available commercial devices and a control law for controlling these devices. Since AC variables are dealt with at the input as well as output side, these SPST switch must be capable of blocking voltages of both polarity and conducting current in both directions.

This book chapter aims to elaborate only on some aspects of design and control of MC from the perspective of low to medium power level applications in the distribution side of power system. Before moving to the details, an overview of 3-phase MC is provided in the following sections to impart an overall understanding of the key design and operational aspects of MC. Those readers who are familiar with MCs may skip this part and start from Sect. 1.2.

1.1 Overview of 3-Phase Matrix Converter

The different design/operational aspects of MC can be broadly be categorized into the following: hardware design, commutation of switches, structure of the topology, modulation and control schemes. These are summarized in the following sections.

1.1.1 Hardware Design

Hardware design begins with the realization of a bidirectional switch. If the switching laws are violated it may lead to short circuit at the input side leading to high input current and eventually failure of the converter. On the other hand if the current flowing through an inductive load is interrupted due to error in switching it would result in high voltage appearing across the switch and eventually lead to its failure. So a protection mechanism for the switches in these situations is a critical feature in design of the hardware. Another very important aspect in hardware design is designing the switching frequency ripple filters. Each of these aspects are discussed below.

1.1.1.1 Realization of a Bidirectional Switch

A bidirectional switch should be capable of conducting current as well as blocking voltages of both polarity. This cannot be achieved by using a single discrete semiconductor device and thus it has to be configured by using more than one switch.

Different ways of realizing a bidirectional switch with discrete IGBT and diodes are shown in Fig. 1.8. Figure 1.8a shows one of the configurations, which is however not used in MCs. One of the reasons behind this is presence of highest number of devices in the current conduction path, leading to higher conduction losses. A more important reason is that it is not possible to selectively allow or block current of a particular direction through this bidirectional switch. So if an output phase has to be switched from one input phase to another, it cannot be done without turning off this switch which leads to violation of the fundamental switching rules. Readily available anti parallel arrangement of diode and IGBTs can be connected to form either the common emitter (CE) or common collector (CC) configurations shown in Fig. 1.8b, c respectively. In terms of device count, bidirectional switch using Reverse blocking (RB) IGBTs shown in Fig. 1.8d is clearly the best option. However RB IGBTs are still not widely available.

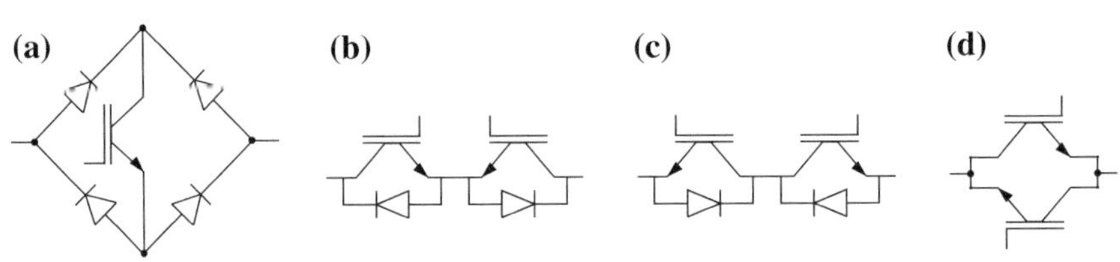

Fig. 1.8 Realizing bidirectional switch. **a** Diode bridge with IGBT, **b** common emitter, **c** common collector, **d** reverse blocking IGBTs

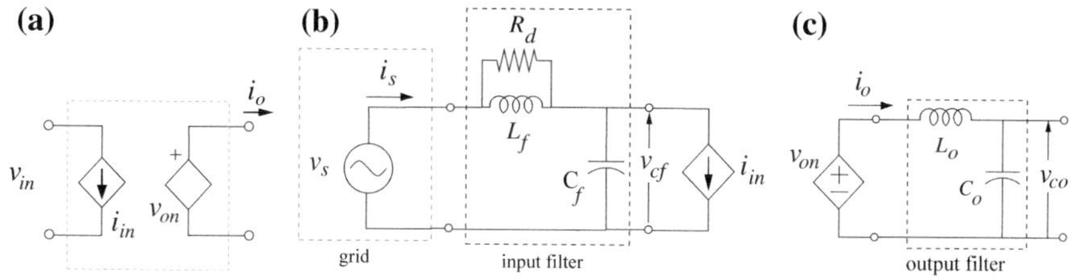

Fig. 1.9 **a** Power stage of MC looking from source and load side. **b** Input filter. **c** Output filter

1.1.1.2 Filter Design

Like a buck converter, the power stage of MC presents current and voltage stiff terminal properties when viewed from the source and load respectively. This is shown in Fig. 1.9a where the MC is modeled as appropriate controlled sources. Balanced 3 phase system has been considered which allows analysis on a single phase basis.

The controlled sources shown in Fig. 1.9a are composed of fundamental as well as switching frequency (f_s) components. Hence, the input side has to be interfaced to the input source through ripple filters to ensure only low-frequency interaction at fundamental frequency. Input filter is an essential requirement of this topology for providing a local circulating path to the switching frequency current. Output side filters are also necessary in applications which require a filtered voltage to be applied to the load. At the same time it has to be ensured through design that addition of filters do not degrade performance parameters like voltage regulation, efficiency and size. These place conflicting demands on the choice of filter elements. Moreover, as the basic structure of MC is made of bi-directional switches, the only associated inertial elements in this topology are the filter components. Consequently the filter parameters affect the system dynamics significantly.

Input filter design has been discussed in [1–6], where, from cost and weight considerations, single stage LC filter has been found to be the most appropriate topology. Figure 1.9b, c show the widely used input and output filter configurations. It is not possible to control the input current in MC, hence a damping resistor is used for the input filters. Although the set of filter elements (L_f, C_f) for a particular resonant frequency is infinite, [2] advocates a maximum C_f to ensure minimum leading Input Displacement Factor (IDF) at low loading conditions. However, in a distribution system most of the loads are inductive and hence this should not be a major concern at lighter loads. An exhaustive treatment of input/output filters with focus on reducing electromagnetic interference and common-mode voltages has been provided in [7]. However, most of these have not investigated the comprehensive design of filters in the context of dynamic performance requirements or reliability of commutation hardware. Filter design, considering stability limits detailed in [8, 9], has been discussed in [6]. However, it will be discussed in a subsequent chapter of this book, that with a proper choice of the modelling paradigm, the derived plant has non-minimum phase zeros but minimum phase poles for all operating points.

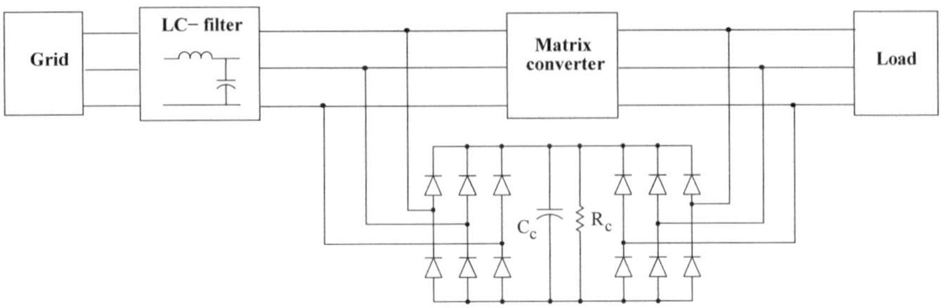

Fig. 1.10 Clamp circuit

Therefore a stable plant is obtained. The relation between filter parameters and non-minimum phase zeros would be detailed and the modifications necessary in filter design will also be discussed.

1.1.1.3 Protection Schemes

Standard protection schemes like overcurrent and overvoltage protection used in any converter are also necessary in MCs. The switches need to be protected in the event of short circuit at input side and open circuit at the output side. Protection for high current during shorting of input terminals can be realized on the gate drivers of the switches themselves. Standard technique like sensing the collector to emitter voltage across an IGBT to detect short circuit can be employed.

Interruption of load current, during communication or due to activation of protection circuit following any fault, may result in high voltage at the output side. Normally a diode clamp circuit along with a capacitor (C_c) and discharging resistor (R_c) as shown in Fig. 1.10 is used to absorb the energy of inductive load [4]. Design of the clamp capacitor has been presented in [5]. Clamp circuits requiring lesser number of diodes have also been proposed in the same paper. This circuit also protects the MC from overvoltages at grid side. Overvoltage protection using varistors and zener diodes with high blocking capability is reported in [10].

1.1.2 Commutation

One of the major hurdles in controlling MCs is switching a load from one input terminal to another without violating the fundamental switching rules. This difficulty in switching can be understood with the example of a two phase to single phase MC feeding an inductive load shown in Fig. 1.11. The bidirectional switch are realized through the CE configuration using discrete IGBTs and diodes. IGBTs in each bidirectional switch (S_1 and S_2) supporting conduction in the forward (towards

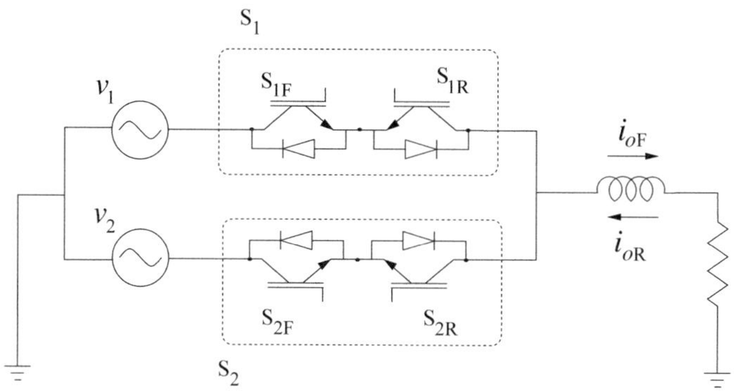

Fig. 1.11 2 phase to 1 phase MC

load) or backward are denoted with additional subscripts F and R respectively. Load was initially connected to v_1 and a switchover to v_2 is required. Now, simultaneous switching of S_1 and S_2 shorts the input terminals, while sequential switching leads to momentary opening of load terminal.

To avoid this a multi-step commutation process based on either output current direction or relative input voltage magnitude [11, 12] becomes necessary. The objectives of these methods are to ensure that the current is not interrupted and at the same time input terminals are not shorted at any instant during the process of commutation. These approaches are detailed in the following subsections.

1.1.2.1 Output Current Direction Based Commutation

Output current direction must be correctly determined for this method to be successful. Figure 1.12a shows the switching sequence from S_1 to S_2 for forward (i_{oF}) and reverse (i_{oR}) direction of i_o, respectively. The sequence is shown in terms of switching function of each IGBT. As an example, commutation sequence for i_{oF} is described. Figure 1.12b shows the different stages of commutation in the two phase to single phase MC. IGBTs shown in bolder mark are those which are turned on (gating pulses are provided), while those which are shown in grey shade indicate the turned off IGBTs.

Step 1. Non conducting IGBT (S_{1R}) in the outgoing bidirectional switch S_1, is first turned off.

Step 2. Next, the IGBT in the incoming S_2 that can support conduction of i_{oF} i.e. S_{2F} is turned on. If $v_2 > v_1$, then natural commutation from S_1 to S_2 takes place at this stage.

Step 3. S_{1F} is turned off subsequently. If $v_1 > v_2$, then forced commutation from S_1 to S_2 occurs at this stage.

Step 4. Finally, S_{2R} is turned on.

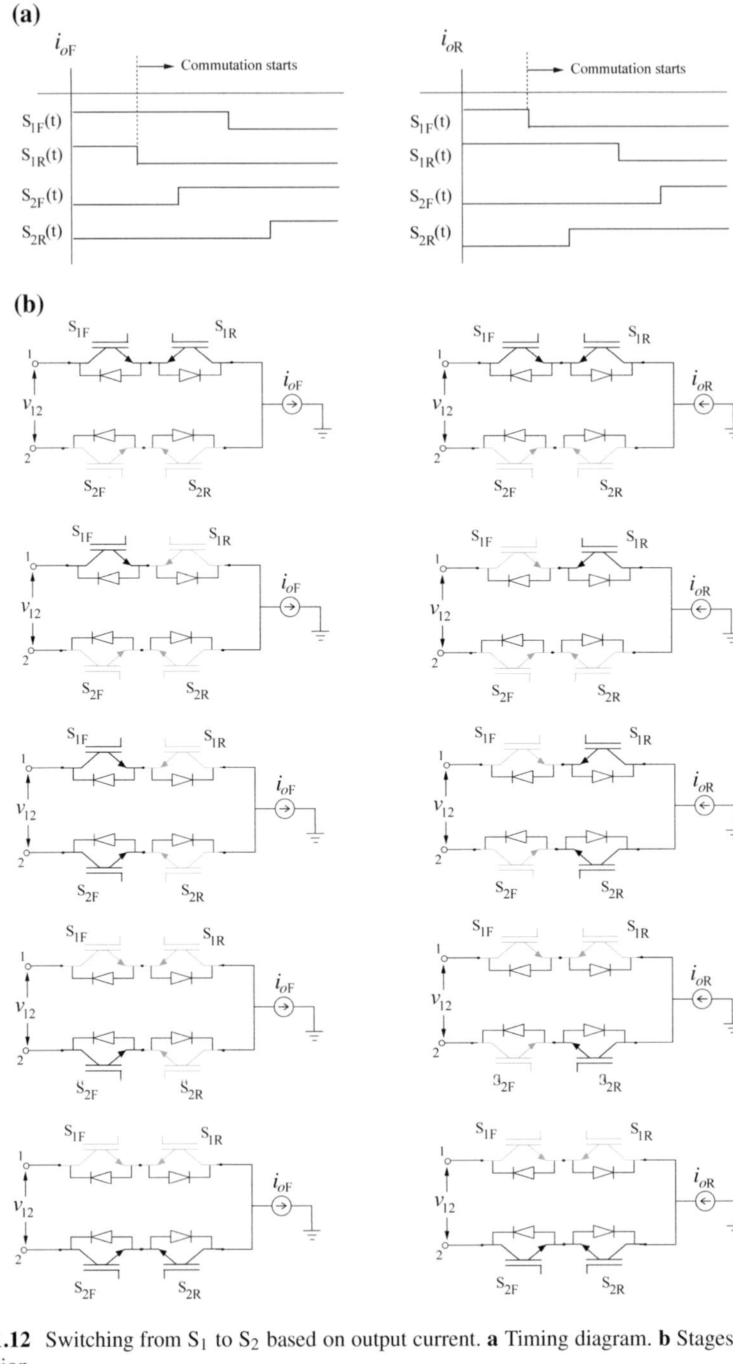

Fig. 1.12 Switching from S_1 to S_2 based on output current. **a** Timing diagram. **b** Stages of commutation

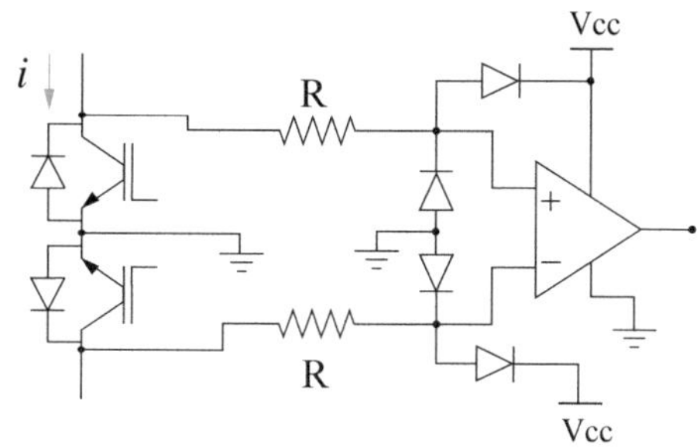

Fig. 1.13 Estimating current direction by sensing the voltage across a bidirectional switch

In steady state, both IGBTs in a bidirectional switch are turned-on to allow reversal of current. The commutation process is thus achieved through 4 steps of turn off-on-off-on sequence. Since turning on times of IGBTs are much lower than the turn off duration, the second and third step can be concurrently executed thereby resulting in a three step commutation method.

Since the first step involves turning off an IGBT, incorrect current direction measurement would lead to open circuit of the load. During the zero crossovers (ZC) of i_o, correctly sensing its direction may become difficult. To overcome this, a method where the direction of i_o is derived by measuring voltage across each device has been reported in [13]. Conceptually this amounts to using an arrangement similar to what is shown in Fig. 1.13. The collector voltage of each device is fed to a comparator, whose output state determines the instantaneous current direction. For example, if the direction of current i is downward as shown, the comparator output would be approximately V_{cc}. Since the device has to block the input line to line voltage in blocking mode, the voltage at input of the comparator has to be clamped to the control voltage level. Hence, the major fraction of the voltage appears across R. So the value of R has to be high enough to restrict the power loss to an acceptable level. Apart from the increased number of components in this method, use of a high R along with unavoidable stray capacitances makes the measurement prone to errors due to the inherent delays involved [14, 15].

1.1.2.2 Input Voltage Magnitude Based Commutation

This method requires knowledge of the relative magnitudes of input voltages. Figure 1.14a, b shows the switching sequence from S_1 to S_2 for different polarities of $v_1 - v_2$. For $v_1 > v_2$, the commutation process as shown in these figures is achieved in the following manner.

Step 1. S_{2F} is turned on. This activates the conduction path for i_{oF} in S_2, while ensuring that there is no circulation current between the two input phases.

Step 2. S_{1F} is turned off. At this instant, if i_o is in forward direction ($i_o = i_{oF}$), conduction is taken up by S_2 i.e. forced commutation takes place here. However, if i_o is in the backward direction ($i_o = i_{oR}$) conduction would still be supported by S_1.

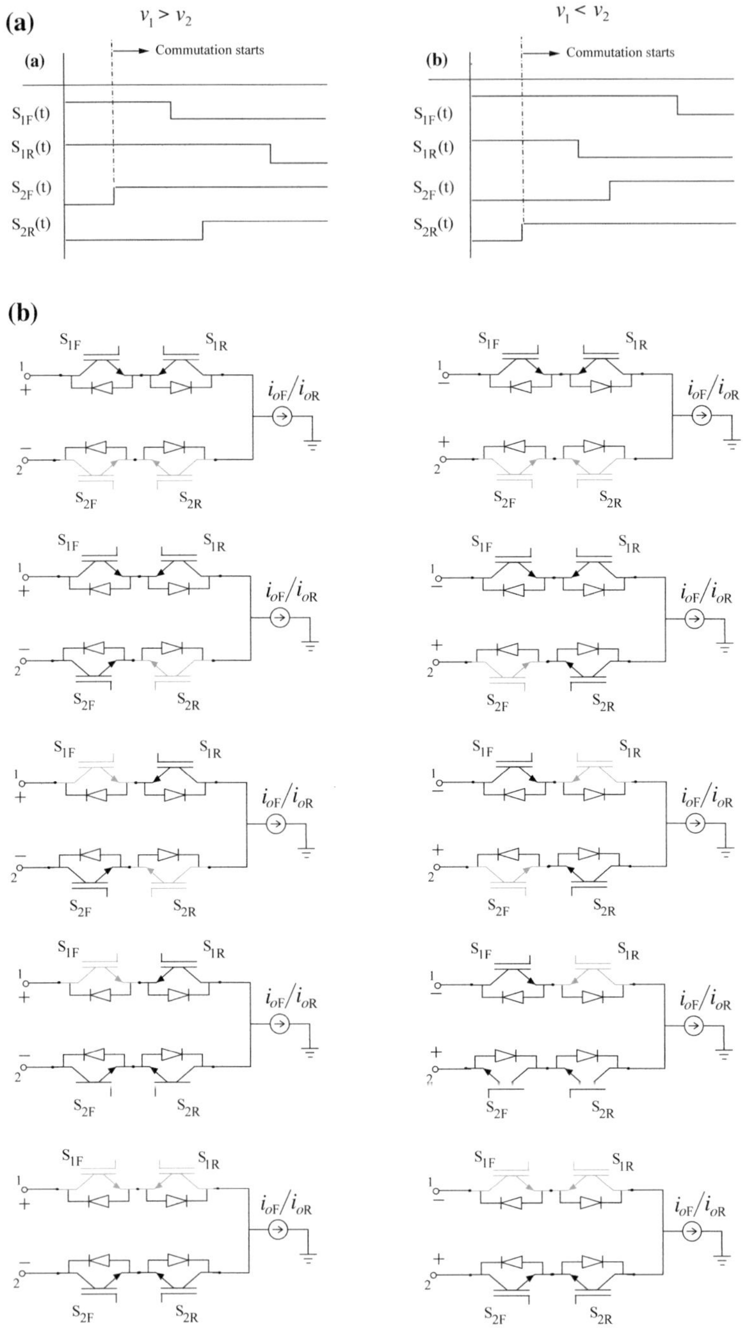

Fig. 1.14 Switching from S_1 to S_2 based on relative magnitude of v_1 and v_2. **a** Timing diagram. **b** Stages of commutation

Step 3. S_{2R} is turned on. Therefore conduction path for i_{oR} gets activated in S_2 and since $v_1 > v_2$, natural commutation occurs for reverse current flow.

Step 4. Finally S_{2F} is turned off.

If the relative voltage magnitudes can be clearly distinguished the first two steps and the third and fourth step can be merged to form a two step sequence.

Voltage based commutation starts with a turning on process. Here, the major concern is correct polarity detection of phase to phase input voltage during its ZC. Falsely detecting this polarity would lead to shorting of the two input terminals. The difficulty in detecting the polarity is compounded by the presence of switching frequency ripple component. Around ZC of the line to line voltage the magnitude of its fundamental component is small but the same does not hold for the ripple component. Also it is much more difficult to accurately measure the switching frequency component than the slowly varying fundamental component.

This issue has been addressed in various papers on commutation. In [16], an analysis of critical window width around ZC of input line voltages has been provided, based on the magnitude of fundamental component of input voltage and resonant component that appears due to the input filter oscillations. During commutation between two phase voltages lying in this defined window, an intermediate step is introduced where the switch over takes place through the third phase. However the effect of the input switched current which plays the major role behind switching frequency ripple in input voltage has been ignored.

In [17], it is shown that with proper zero vector placement for Space vector modulation (SVM) based approaches, safe commutation can be achieved in spite of voltage measurement inaccuracies. This method is restricted to operation within a certain input displacement angle ($\pm\theta°$) around voltage ZC, where θ is to be decided on the basis of the specific input ripple voltage measurements for a given hardware. Moreover, as this method depends on minimum duration of zero vectors it is not applicable to all operating points. For all operating regions and in applications requiring wider control of IDF [18, 19], commutation at ZC of line-line voltage is still difficult, particularly at high output current amplitude.

There is no guide for conclusively selecting a particular method for an application. However, applications having a high inductance at the output side will have a low switching ripple component in the output current. Therefore it is easier to sense to measure the output current correctly and thus current based techniques may be better suited for such conditions. For applications requiring a regulated sinusoidal output voltage, an output filter becomes necessary. The inductor in this filter which is in series with the output terminals must be small so that the voltage drop is less. Consequently the ripple components of the current would be large and so would be the magnitude of error in measuring it. The applications considered in this work falls in the latter category. Existing input voltage based commutation method is adopted in this work. A closed-form expression of the ripple voltage in the input filter capacitor and error in measurement has been derived. This is used to derive the minimum size of filter capacitor as would be detailed in a subsequent chapter.

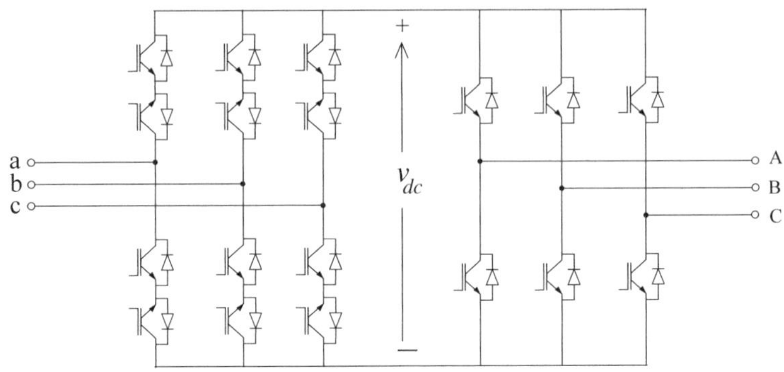

Fig. 1.15 Indirect MC

1.1.3 Topology

The structure shown in Fig. 1.7 is referred to as the Direct Matrix Converter (DMC). The operational challenges associated with DMC have prompted engineers to find out variants of this basic structure that may simplify some of the operational issues.

Figure 1.15 shows a variant of the basic MC topology. This is known as Indirect MC (IMC) [20]. The power processing in this structure is achieved through rectification and inversion process as found in B2B VSIs. Commutation is reported to be simpler than DMCs [20, 21], as the rectifier stage bidirectional switches can be switched when the inverter side is in free wheeling mode. In this condition, commutation at the rectifier side is not constrained by any restriction on the opening of the output terminals and only needs to ensure that the input terminals are never shorted. Since the DC link voltage (v_{dc}) has to be positive, IDF can be controlled in the range 0.866–1. Reduction of IMC structure leading to varying forms of 'Sparse Matrix Converters' has been summarized in [21].

This work is related to power system applications for synchronous systems. In such applications reactive power plays a significant role. Although control of input reactive power has not been attempted in this work, it can be included in future investigations by extending the analysis presented here. Since DMC offers a much wider scope of IDF control than IMC, it has been adopted here. Hence, subsequent discussions are related to DMCs. However since the basic functional feature of both DMC and IMCs are same, discussion related to dynamic model, filter design and modulation applies equally to both topologies.

1.1.4 Modulation

A large number of modulation techniques have been reported, dating back to the inception of MC. A few of those, which have been critically evaluated, are discussed here.

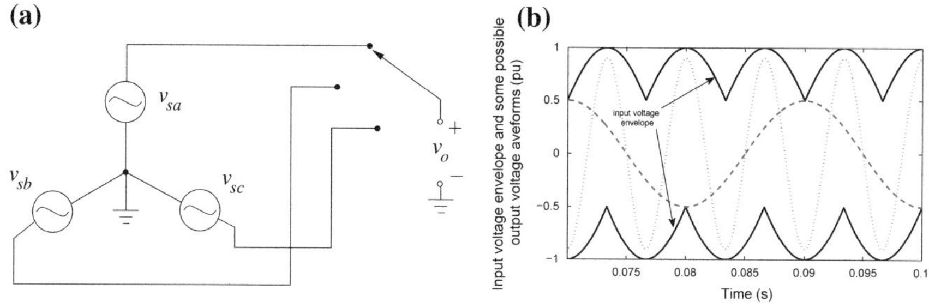

Fig. 1.16 **a** Input voltage. **b** Output voltage envelope for all 27 switching combinations

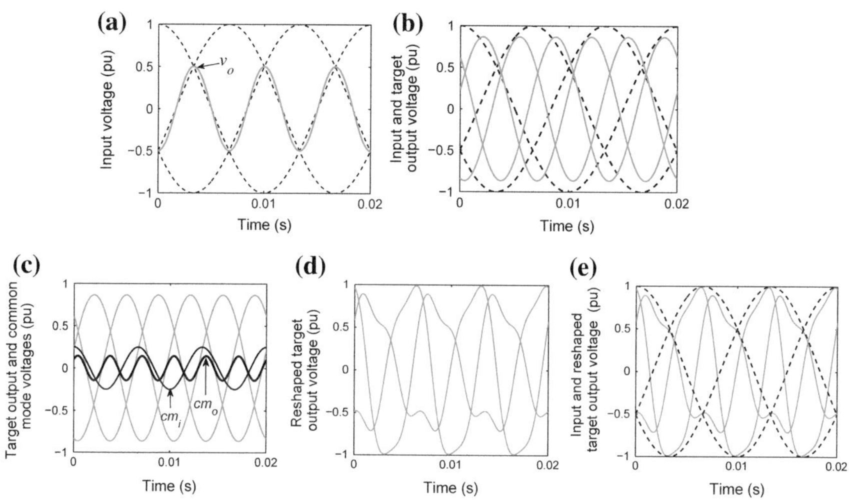

Fig. 1.17 **a** Input voltages. **b** Input and target output voltages. **c** Target output voltages along with common mode voltages. **d** Reshaped target voltages. **e** Input along with reshaped target output voltages

1.1.4.1 Direct Transfer Function Based Approach

One of the earliest papers [22] known to provide a rigorous analysis of a 3-phase MC demonstrated this approach (Fig. 1.16).

This method derives the low frequency modulation signals directly from the reference and input voltages. In the first algorithm [22] the envelope of output voltage waveform to be synthesized must remain within the input voltage envelope at all instants of time. Consequently, as shown in Fig. 1.17a, the maximum achievable output voltage (v_o) amplitude is 0.5 times of the input. The performance was significantly improved in the second algorithm [23] where the maximum 0.866 voltage gain ratio could be reached. The essence of the second algorithm is explained with the help of Fig. 1.17b–e. Figure 1.17b shows the input and target output voltages whose amplitude are 0.866 times of the input. The output waveform is observed to exceed the input waveform envelope at some intervals. By appropriately adding common mode (cm) terms, the second algorithm reshapes the target voltage envelope to completely fit into the input envelope. Figure 1.17c shows the third harmonic cm components for input (cm_i) and output (cm_o) which are added on to the target output waveform. Figure 1.17d shows the reshaped target waveforms, the envelope of which is now completely enclosed by that of the input waveform, as shown in Fig. 1.17e. Since the load and supply neutrals are usually isolated and the cm terms disappear from the line to line voltages, addition of these low frequency terms are justified. Denoting the input and target output voltage amplitude and frequency as $\hat{V}_i$, $\hat{V}_o^*$, and ω_i, ω_o respectively, the described procedure is equivalent to generating the output voltage references as [23],

$$\hat{V}_o^R\begin{bmatrix}\cos(\omega_o t)\\ \cos(\omega_o t-120^\circ)\\ \cos(\omega_o t+120^\circ)\end{bmatrix}=\hat{V}_o^*\begin{bmatrix}\cos(\omega_o t)\\ \cos(\omega_o t-120^\circ)\\ \cos(\omega_o t+120^\circ)\end{bmatrix}+\underbrace{\frac{\hat{V}_i}{4}\begin{bmatrix}\cos(3\omega_i t)\\ \cos(3\omega_i t)\\ \cos(3\omega_i t)\end{bmatrix}}_{cm_i}-\underbrace{\frac{\hat{V}_o^*}{6}\begin{bmatrix}\cos(3\omega_o t)\\ \cos(3\omega_o t)\\ \cos(3\omega_o t)\end{bmatrix}}_{cm_o} \quad (1.4)$$

If the output load angle (ϕ_L) can be measured, IDF can also be controlled using the second algorithm. The same maximum voltage gain and IDF control can be achieved by other modulation methods without the necessity of measuring ϕ_L and therefore having a relatively simple hardware realization. They are described in the following section.

1.1.4.2 Space Vector Modulation (SVM)

The power stage of MC can be viewed [24] as a cascaded rectifier-inverter stage coupled by a fictitious DC link as shown in Fig. 1.18. SVM based on this decoupled construct allowing control of IDF has been described in [25]. This method is referred

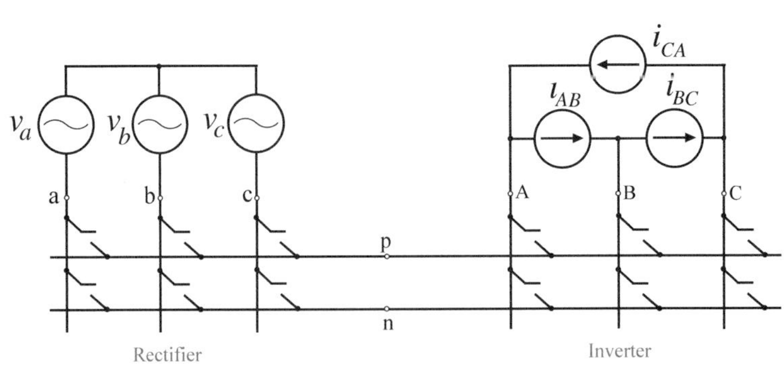

Fig. 1.18 Decoupled rectifier-inverter construct

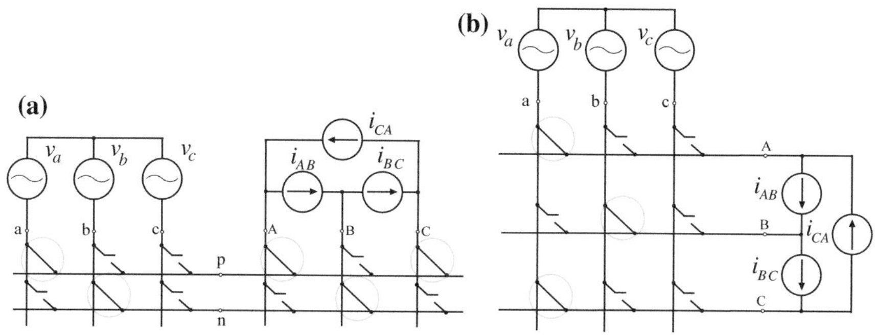

Fig. 1.19 Mapping the switching functions

in this book as Indirect space vector modulation (ISVM). Here, the duty ratios of the switches for individual stages are initially calculated and then combined on the basis of instantaneous active power balance between input and output. Thereafter, the switching functions for the twelve switches are mapped to the nine 4Qsws MC structure. An example of this mapping from the 12 to 9 switch structure is shown in Fig. 1.19a, b. Suppose the input terminal a in the decoupled structure needs to be connected to output terminals A and C, while terminal b to output terminal B as shown in Fig. 1.19a. The corresponding switches which need to be closed in the 9 switch structure are shown in Fig. 1.19b.

A maximum voltage gain of 0.866 for unity IDF is attainable with ISVM. Hardware realization of ISVM is simpler than direct approaches and knowledge of ϕ_L is not required for IDF control. Moreover, space vector based schemes offers substantial freedom in choosing the switching sequence which can be used for reducing common mode voltage at the output [26], output current ripple [27], reducing switching losses [28] etc. These features make SVM one of the most widely adopted modulation method in MCs. The basic working principle of ISVM has been briefly discussed later in this book.

1.1.4.3 Other Pulse Width Modulation (PWM) Based Methods

One of the often stated disadvantages of SVM lies in the use of look-up tables for hardware realization. This has been the primary inspiration for alternate PWM methods like [29] where *cm* terms are added to modulation signals in order to achieve a voltage gain of 0.866. Control of IDF is also possible in this scheme. Another carrier based method [30] emulates the input line to line voltages as the DC link voltage of an inverter and by doing so tries to extend modulation strategies for VSI to MC. The method proposes to use two input line to line voltage as the imaginary DC link voltage over a switching cycle. The input voltages are denoted as V_{max}, V_{med} and V_{min} based upon their relative magnitudes, as shown in Fig. 1.20. The switching period T_s is divided into two parts T_1 and T_2 depending on the input voltage angle.

Fig. 1.20 Describing the discontinuous carrier signals [30]

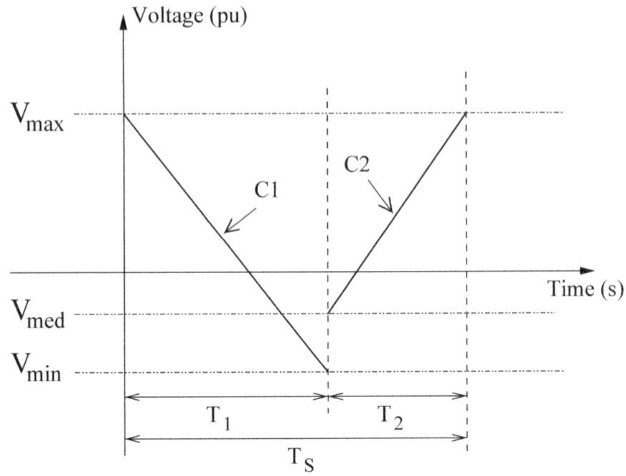

Now for the two separate parts of the switching cycle, two carriers (C_1 and C_2) are defined as the difference between the values of input voltages. The argument behind using two carriers, is, utilization of all three input voltages during a switching period [30]. If two inputs are only used, one of the input phase currents would remain zero thereby leading to higher input current distortion. A voltage gain of 0.866 in the linear modulation region and unity IDF is ensured with this method.

The PWM based methods are gradually becoming an active area of research. The performance in terms of voltage gain and controllable IDF has been proven to be equivalent to SVM methods. However the superiority or equivalence of these in context of the flexibility shown by SVM schemes in choosing and placing the zero vectors is yet to be established.

1.1.4.4 Discrete Methods Based on Predictive Model

Modulation methods based on discrete models have been a very actively researched topic in the last decade. These methods use discrete-time models of the load, filter and converter to predict behaviour of variables like load current, reactive power etc. For example, if the objective is to control the output current (i_o), its value at $k+1$th sampling instant is first computed using the predictive models for all valid switching states. The resulting absolute error between the reference and predicted values for a single valid switching state can be represented as

$$\Delta i_o(k+1) = \left| i_o^*(k+1) - i_o^p(k+1) \right|, \tag{1.5}$$

where * and ^{p} denote reference and predicted values respectively. Subsequently, the state that gives minimum $\Delta i_o(k+1)$ is chosen as the $k+1$th switching state. However, as no current feedback is used, this may cause unchecked deviation of actual current. The strength of the method lies in its flexibility to address multiple control objectives. An example of this was reported in [31], where the objective was to control i_o and minimize input reactive power q_s. A quality function (Q_G) was defined as

$$Q_G = W1 \cdot \Delta i_o(K+1) + W2 \cdot \Delta q_s(K+1), \tag{1.6}$$

where $W1$, $W2$ are weighting functions. An analytical approach of choosing the weighting functions is yet to be reported. On basis of a particular set of $W1$ and $W2$, Q_G is calculated for all the 27 valid switching states of MC. Thereafter the state resulting in minimum Q_G is selected at the next switching instant. Applications of this predictive model based approach for different control objectives are summarized in [32].

Keeping aside the obvious concern regarding the accuracy of the predictive models, on-line computation of Q_G for all valid switching states would increase the computational burden to a great extent. Superiority of this technique over the performance obtained with SVM approaches has not been established yet [32].

1.1.5 Dynamic Model for Controller Design

Tight coupling between the input and output side of MC complicates controller design, particularly for applications requiring high bandwidth. Hence, for stable operation, a proper control scheme based on a tractable dynamic model becomes necessary—an area, which has not received sufficient attention. This has been acknowledged in [33] which presents a dynamic model for MC used as an interface between high speed micro turbine generator and the utility grid. The model is developed using state matrices, which is subsequently utilized for design of active and reactive power control. State feedback approaches based on eigenvalue analysis has been used in [34] for voltage sag mitigation, another application which demands high dynamic performance. With available tools like MATLAB control toolbox, designing controllers using state matrices is not difficult and desired dynamic performance is guaranteed provided the system is well modeled. But these methods provide little insight into the system as associating the eigenvalues, or critical dynamic behavior, with specific physical elements becomes increasingly difficult with higher order systems.

In the context of stable operation of MCs, [8, 9, 35–38] have reported and analyzed existence of a throughput power limit (P_{lim}) for stable operation. In all these studies, for output voltage control, the reference $\bar{V}_o^*$ is obtained treating the output filter and the load as plant. Subsequently the modulation signals are essentially computed as

$$\bar{m} = \frac{\bar{V}_o^*}{\bar{V}_{in}}, \tag{1.7}$$

where $\bar{V}_{in}$ is voltage vector of the input voltage v_{in} (v_{cf}) applied at the input terminals shown in Fig. 1.9a, b. Although $\bar{m}$ is not extracted exactly through a scalar division as it appears in (1.7), the conceptual basis is essentially same. This method is sometimes referred as feed forward compensation of input voltages [8]. P_{lim} is prescribed by investigating the input side of MC assuming constant active power transfer through it. Investigation of how filtering $\bar{V}_{in}$ affects the system eigenvalues has been reported in [36]. Examination of how sampling delays in measuring $\bar{V}_{in}$ affects P_{lim} has been detailed in [8, 9] considering constant output power. More recently [38] has proposed use of the source voltage i.e. v_s in Fig. 1.9b in place of v_{cf} which improves the stability limit for grids having a low source impedance.

A detailed discussion on the dynamic model and appropriate controller design has been provided later in this book chapter.

1.2 Motivation and Objectives

Out of the different research tracks highlighted in the last section, major effort has been spent in investigating modulation algorithms that can extract the optimal performance out of MC in terms of voltage gain, IDF control, relative ease of hardware realization, switching losses etc. This has been followed by devising reliable commutation and protection strategies, alternative topologies with reduced number of switches in certain cases where unidirectional flow of power is required, input filter design etc. One of the lesser investigated area has been the development of a dynamic model that provides a physical insight into the system.

From the applications perspective, although MC is perceived as an alternative to B2B VSI (DAB), research efforts has been largely confined to motor/drive. The last decade have witnessed an emerging academic interest in using MC for power system applications. For example, it has been reported to be used as a reactive power compensator [18, 19], voltage sag compensator [34], regulated utility power supply [39], Unified power flow controller [40] to name a few.

A major motivation of writing this book is to introduce to the reader the considerations for using the MC for synchronous power system applications. Two typical applications were chosen for investigation—regulated voltage supply and voltage sag mitigation. Of course, several research publications have reported MC based solutions, along with dynamic model and control techniques, for interfacing with utility power systems. Section 1.1.5 of this book shows that some of these approaches, which are either based on feed forward compensation of input voltages or on state feedback based schemes, have specific limitations. Some reports, e.g. [34], incorporate an additional flywheel energy storage which negates the high power density advantages of the MC. Hence, the broad aim of this book is categorized into the following objectives.

1. Complexity of control has been widely acknowledged to be one of the major reasons for the low industrial acceptance of MC, even after three decades of

intensive research. Control may become more difficult as one moves out of the motor/drives domain to power system applications, which demand faster dynamic performance. For a tightly coupled input-output unit like the MC, the system designer cannot take an independent approach towards design of sub components like filter and controller, without addressing common constraints that bind both. A dynamic model that helps to identify such common links and also aid the physical understanding of the system is presented in this book. So development of a tractable dynamic model by which the stated concerns can be addressed was set to be the first objective of the work.
2. The next goal was to obtain a design guideline for filters which should also take into consideration the constraints that emerge from the dynamic model and sizing of passives for reliable commutation.
3. Once the first two targets were achieved, controller design naturally became the third objective. Given that a recurring argument against using MC has been the difficulty of control, emphasis was given on finding a control scheme that is easily realizable.
4. Experimental validations of the analytical claims was set as the final objective.

1.3 Assumptions and Scope

The analysis presented in this book uses a set of assumptions, which are routinely made in MC analysis but are stated nevertheless for clarity. This also helps to delineate the scope of this book, which are listed below to enable correct evaluation of the material presented here.

- Plant model is derived assuming 3 phase balanced system. Harmonic distortions in supply voltage has not been considered.
- Modelling is confined to synchronous applications.
- Ideal converter and filter elements have been assumed.
- Dynamics of Phase lock loop (PLL) have not been considered.
- Harmonic analysis of MC has not been carried out.
- For voltage sag mitigation a 3 wire linear load is considered thereby removing any possibility of zero sequence component in load.

1.4 Layout of the Book

The rest of the book is structured in the sequence the objectives were described. The next chapter i.e. Chap. 2 describes the development of linearized plant model based on the low frequency gain of MC. The same model is used for the investigation of the plant for non-minimum phase poles/zeros. Chapter 3 presents the filter design approach where the different performance criteria are set in the process to find an acceptable solution set for filter parameters that satisfy all of them. Relevant experimental results are presented. The next two chapters deal with the application aspects where the design techniques have been applied. In Chap. 4, a controller design approach for an application requiring regulated 3 Phase sinusoidal voltage is described. Experimental results are presented. The control design approach is then extended with appropriate modifications in the next chapter where MC is used as a voltage sag mitigating device. Chapter 5 discusses symmetrical and asymmetrical voltage sag mitigation using MC for linear loads. Experimental results are presented. Some critical observations based on the presented material are summarized in Chap. 6 along with some observations of the authors.

Chapter 2
Low Frequency Dynamic Model

Like other converters in the switched mode power converter family, operation of matrix converters involves two significant frequency components. These are the fundamental frequency and the switching frequency. The fundamental frequency component is associated with power transfer while the switching component is a byproduct of the operation. If these switching components are injected into the grid side or to the load side it would lead to undesirable consequences and thus it is necessary to filter them. The filtering process has been discussed in detail in a later chapter. For this chapter, it is sufficient to have the understanding that low pass filters are realized by passive elements in these converters. For a given power level, the size of these filter components reduces as the switching frequency becomes higher. Hence a large difference between the switching frequency and the fundamental component is always preferable as it helps to achieve effective filtering with smaller components size. The switching frequencies are in the order kHzs, or tens of kHzs, a number which depends upon the rated power handling capacity of the converter. On the other hand, the order of the fundamental component is usually tens of Hertz. For any well designed converter, the switching frequency component in the output variables are negligible compared to the fundamental frequency component. Therefore, while deriving a model which represents the dynamic behaviour of the converter the high switching frequency component can be neglected without any loss of accuracy in modelling. Thus the control scheme of MC or for many other SMPCs has to be devised such that the fundamental frequency component and low frequency harmonic components can be processed with a desired degree of accuracy.

This chapter presents an analysis of the linearized low frequency model of MC, where it has been used as a 3-phase regulated sinusoidal voltage supply. The objectives of all modulation techniques for MC are attaining the desired—amplitude, phase and frequency of the output voltages and IDF at the input side. In the first section of this chapter, the modulation technique adopted in this work is briefly discussed to describe the low frequency gain of MC. It has been subsequently used to develop the linearized dynamic model in Sect. 2.2 from where the plant transfer function matrix

A. Dasgupta and P. Sensarma, *Design and Control of Matrix Converters*,
Energy Systems in Electrical Engineering, DOI 10.1007/978-981-10-3831-0_2

is derived. In Sect. 2.3 it is demonstrated how the system poles/zeros are affected by input voltage, input displacement factor, throughput power and parameters of the input filter. In the process it is found that based on the operating points, right half zeros may emerge in the control to output transfer functions beyond a critical level of input power.

2.1 Low Frequency Gain of 3-Phase MC

The phase to input side neutral voltages applied to the converter i.e. the voltages at terminals a, b and c with reference to the input neutral shown in Fig. 2.1 are represented as

$$\begin{bmatrix} v_a \\ v_b \\ v_c \end{bmatrix} = \mathbf{v_{cf}}^{(3)} = \hat{V}_{cf} \begin{bmatrix} \cos(\omega_i t - \varphi_c) \\ \cos(\omega_i t - \varphi_c - 120^\circ) \\ \cos(\omega_i t - \varphi_c + 120^\circ) \end{bmatrix}. \tag{2.1}$$

These voltages appear across the input filter capacitor C_f depicted in the per-phase diagram shown in Fig. 1.9b and reproduced in Fig. 2.2. MC allows control on the magnitude, frequency and phase of the output voltage and IDF. Let the desired output voltages, referred to output neutral n, be described as

$$\mathbf{v^*_{on}}^{(3)} = \begin{bmatrix} v^*_{An} \\ v^*_{Bn} \\ v^*_{Cn} \end{bmatrix} = \hat{V}_o \begin{bmatrix} \cos(\omega_o t + \varphi_o) \\ \cos(\omega_o t + \varphi_o - 120^\circ) \\ \cos(\omega_o t + \varphi_o + 120^\circ) \end{bmatrix}, \tag{2.2}$$

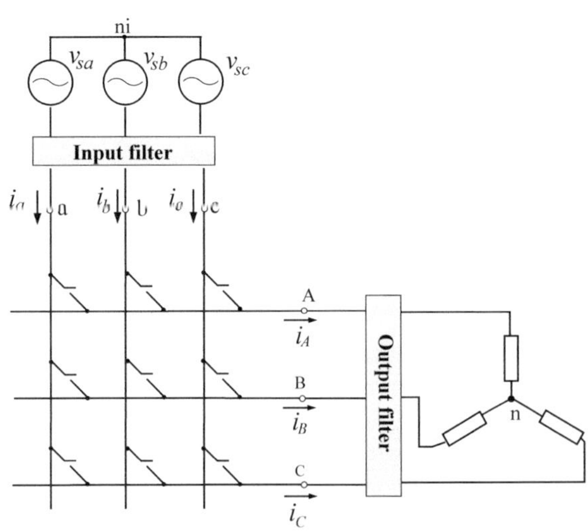

Fig. 2.1 Schematic of a 3-ph MC

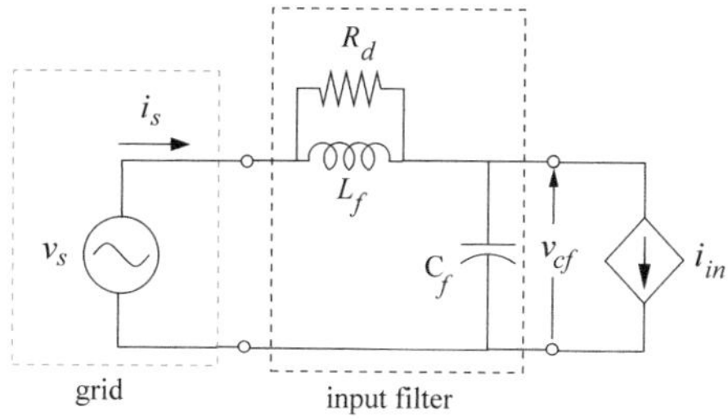

Fig. 2.2 Input side equivalent circuit for the per-phase system

where * indicates reference signal. The corresponding line to line voltages are

$$\mathbf{v^*}_{\mathbf{oL}}^{(3)} = \begin{bmatrix} v^*_{An} - v^*_{Bn} \\ v^*_{Bn} - v^*_{Cn} \\ v^*_{Cn} - v^*_{An} \end{bmatrix} = \begin{bmatrix} v^*_{AB} \\ v^*_{BC} \\ v^*_{CA} \end{bmatrix} = \sqrt{3}\hat{V}_o \begin{bmatrix} \cos(\omega_o t + \varphi_o + 30^\circ) \\ \cos(\omega_o t + \varphi_o + 30^\circ - 120^\circ) \\ \cos(\omega_o t + \varphi_o + 30^\circ + 120^\circ) \end{bmatrix}. \tag{2.3}$$

Indirect space vector modulation (ISVM) [25] method has been adopted in this work to synthesize the desired voltage output. The basic principle of this method is discussed in the following section.

2.1.1 *Indirect Space Vector Modulation (ISVM) Approach*

In ISVM technique the 9 bidirectional switch MC structure is conceptually decoupled into an equivalent cascaded chain of fictitious current and voltage source converters (CSC and VSC) as shown in Fig. 2.3. The imaginary DC link voltage and current i.e. V_{dc} and I_{dc} are treated as sources for the VSC and CSC parts respectively. The duty cycle of each switch in the 9 bidirectional switch MC structure is calculated in two stages. In the first stage, the duty cycles are calculated separately for each converters of the decoupled construct using conventional space vector modulation (SVM)

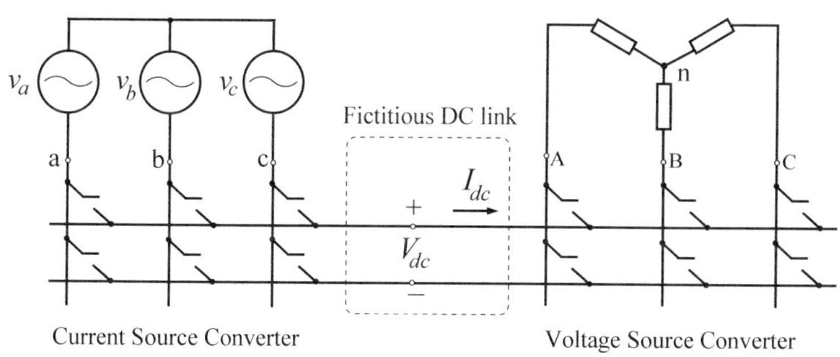

Fig. 2.3 Decoupled current and voltage source converter construct

approach. In the next stage, the individual duty cycles are combined based on instantaneous active power balance between the input and output sides. The duty cycle of the 12 switch fictitious CSC–VSC structure can be then directly mapped to the 9 bidirectional switch MC topology.

To illustrate this procedure, the VSC part is considered first, where the reference line to line voltages are used to define the vector

$$\bar{\mathbf{V}}_{\mathbf{oL}}^{*} = \sqrt{\frac{2}{3}}\left(v_{AB}^{*} + v_{BC}^{*}\, e^{j2\pi/3} + v_{CA}^{*}\, e^{-j2\pi/3}\right). \tag{2.4}$$

Here the constant multiplier $\sqrt{2/3}$ ensures power invariant transformations. Substituting the line voltages from (2.3) in (2.4), $\bar{\mathbf{V}}_{\mathbf{oL}}^{*}$ is obtained as

$$\bar{\mathbf{V}}_{\mathbf{oL}}^{*} = \frac{3}{\sqrt{2}}\,\hat{V}_o\, e^{j(\omega_o t + \varphi_o + 30^\circ)}. \tag{2.5}$$

Referring to Fig. 2.3, there are 8 valid switching combinations by which the output terminals of the VSC can be connected to the fictitious DC link. Here, the term valid combinations refers to those which ensure that any output terminal is never left open. The line voltages corresponding to each of these combinations are used to create the vector

$$\bar{\mathbf{V}}_{\mathbf{oL}} = \sqrt{\frac{2}{3}}\left(v_{AB} + v_{BC}\, e^{j2\pi/3} + v_{CA}\, e^{-j2\pi/3}\right). \tag{2.6}$$

Applying (2.6) to each of the 8 combinations result in stationary vectors (SV) which are sometimes also referred to as switching state vectors. These are tabulated in Table 2.1 and also shown in Fig. 2.4a.

Table 2.1 Switching state vectors for the VSC

A	B	C	$\bar{\mathbf{V}}_{\mathbf{oL}}$	Stationary vectors
+	−	−	$\sqrt{\frac{2}{3}}\,[V_{dc} + 0 - V_{dc}\, e^{-j2\pi/3}] = \sqrt{2}V_{dc}\angle 30^\circ$	$\mathbf{V}_1$
+	+	−	$\sqrt{\frac{2}{3}}\,[0 + V_{dc}\, e^{j2\pi/3} - V_{dc}\, e^{-j2\pi/3}] = \sqrt{2}V_{dc}\angle 90^\circ$	$\mathbf{V}_2$
−	+	−	$\sqrt{\frac{2}{3}}\,[-V_{dc} + V_{dc}\, e^{j2\pi/3} + 0] = \sqrt{2}V_{dc}\angle 150^\circ$	$\mathbf{V}_3$
−	+	+	$\sqrt{\frac{2}{3}}\,[-V_{dc} + 0 + V_{dc}\, e^{-j2\pi/3}] = \sqrt{2}V_{dc}\angle 210^\circ$	$\mathbf{V}_4$
−	−	+	$\sqrt{\frac{2}{3}}\,[0 + -V_{pn}\, e^{j2\pi/3} + V_{pn}\, e^{-j2\pi/3}] = \sqrt{2}V_{dc}\angle 270^\circ$	$\mathbf{V}_5$
+	−	+	$\sqrt{\frac{2}{3}}\,[V_{pn} - V_{pn}\, e^{j2\pi/3} + 0] = \sqrt{2}V_{dc}\angle 330^\circ$	$\mathbf{V}_6$
−	−	−	$\sqrt{\frac{2}{3}}\,[0 + 0 + 0] = 0$	$\mathbf{V}_0$
+	+	+		

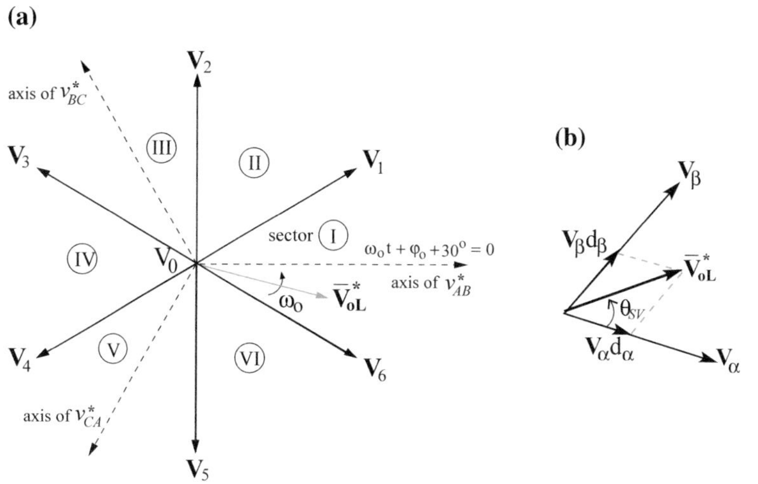

Fig. 2.4 **a** Output voltage stationary vectors and $\bar{\mathbf{V}}^*_{\mathbf{oL}}$. **b** Synthesizing $\bar{\mathbf{V}}^*_{\mathbf{oL}}$

It is evident that there are 6 sets of SVs with non zero magnitude having distinct spatial orientation and 2 sets of SVs with zero magnitude. The objective of SVM is to synthesize $\bar{\mathbf{V}}^*_{\mathbf{oL}}$, in a time-averaged sense, over a switching period (T_S), by using the available SVs as

$$\int_0^{T_S} \bar{\mathbf{V}}^*_{\mathbf{oL}} dt = \int_0^{T_S} \bar{\mathbf{V}}_{\mathbf{oL}} dt. \tag{2.7}$$

Since the switching frequency (f_S) is much higher than the fundamental frequency, $\bar{\mathbf{V}}^*_{\mathbf{oL}}$ practically behaves like a stationary vector over a single switching period. So with appropriate scaling, many combinations of SVs can be used to synthesize the target voltage vector $\bar{\mathbf{V}}^*_{\mathbf{oL}}$. However, using two adjacent SVs leads to minimum number of switching. Additionally, the instantaneous voltage deviation $\bar{\mathbf{V}}^*_{\mathbf{oL}} - \bar{\mathbf{V}}_{\mathbf{oL}}$ is lowest, which ensures minimum current ripple in the inductors connected to the output terminals. The adjacent SVs, $\mathbf{V}_\alpha$ and $\mathbf{V}_\beta$, are shown in Fig. 2.4b trailing and leading $\bar{\mathbf{V}}^*_{\mathbf{oL}}$ respectively. θ_{SV} is the instantaneous angle between $\bar{\mathbf{V}}^*_{\mathbf{oL}}$ and $\mathbf{V}_\alpha$. Over a single switching period T_S, assumption of a stationary $\bar{\mathbf{V}}^*_{\mathbf{oL}}$ implies a constant θ_{SV}. Therefore, $\bar{\mathbf{V}}^*_{\mathbf{oL}}$ can be synthesized as

$$\begin{aligned} \bar{\mathbf{V}}^*_{\mathbf{oL}} &= \mathbf{V}_\alpha \frac{T_\alpha}{T_S} + \mathbf{V}_\beta \frac{T_\beta}{T_S} \\ &= \mathbf{V}_\alpha d_\alpha + \mathbf{V}_\beta d_\beta. \end{aligned} \tag{2.8}$$

Applying properties of triangle,

$$\frac{|\bar{\mathbf{V}}^*_{\mathbf{oL}}|}{\sin 120^\circ} = \frac{|\mathbf{V}_\alpha d_\alpha|}{\sin(60^\circ - \theta_{SV})} = \frac{|\mathbf{V}_\beta d_\beta|}{\sin(\theta_{SV})}. \tag{2.9}$$

Table 2.2 Line voltages, when $\bar{\mathbf{V}}^*_{\mathbf{oL}}$ is in sector I

	ON time	A	B	C		v_{AB}	v_{BC}	v_{CA}
During $d_6\,(d_\alpha)$	$d_\alpha T_S$	+	–	+	$\longrightarrow$	V_{dc}	$-V_{dc}$	0
During $d_1\,(d_\beta)$	$d_\beta T_S$	+	–	–	$\longrightarrow$	V_{dc}	0	$-V_{dc}$

Using (2.5), Table 2.1 and (2.9), d_α and d_β are determined as

$$\begin{aligned} d_\alpha &= \frac{\sqrt{3}\,\hat{V}_o}{V_{dc}}\sin(60^\circ - \theta_{SV}) = m_v \sin(60^\circ - \theta_{SV}), \\ d_\beta &= \frac{\sqrt{3}\,\hat{V}_o}{V_{dc}}\sin(\theta_{SV}) = m_v \sin(\theta_{SV}). \end{aligned} \tag{2.10}$$

As an example, let $\bar{\mathbf{V}}^*_{\mathbf{oL}}$ be considered to be lying in sector I. Here $\mathbf{V}_\alpha = \mathbf{V}_6$ and $\mathbf{V}_\beta = \mathbf{V}_1$. Using Table 2.1, the line voltages corresponding to the two duty cycles are shown in Table 2.2.

Using (2.10) and Table 2.2, the average value of the synthesized output voltages over T_S, when $\bar{\mathbf{V}}_{\mathbf{oL}}$ is in sector I, are obtained as

$$\mathbf{v_{oL}}^{(3)} = \begin{bmatrix} v_{AB} \\ v_{BC} \\ v_{CA} \end{bmatrix} = \begin{bmatrix} d_\alpha + d_\beta \\ -d_\alpha \\ -d_\beta \end{bmatrix} V_{dc} = m_v \begin{bmatrix} \cos(\theta_{SV} - 30^\circ) \\ -\sin(60^\circ - \theta_{SV}) \\ -\sin(\theta_{SV}) \end{bmatrix} V_{dc}. \tag{2.11}$$

Again from Fig. 2.4a, b, when $\bar{\mathbf{V}}_{\mathbf{oL}}$ is in sector I, θ_{SV} can be evaluated in terms of output voltage frequency and phase as

$$\theta_{SV} - (\omega_o t + \varphi_0 + 30^\circ) = 30^\circ \implies \theta_{SV} = \omega_o t + \varphi_o + 60^\circ. \tag{2.12}$$

Substituting θ_{SV} from (2.12) in (2.11), the output voltages can be expressed as

$$\mathbf{v_{oL}}^{(3)} = \underbrace{m_v \begin{bmatrix} \cos(\omega_o t + \varphi_o + 30^\circ) \\ \cos(\omega_o t + \varphi_o + 30^\circ - 120^\circ) \\ \cos(\omega_o t + \varphi_o + 30^\circ + 120^\circ) \end{bmatrix}}_{\mathbf{m_v}} V_{dc} = \mathbf{m_v} V_{dc}, \tag{2.13}$$

where $\mathbf{m_v}$ represents the low frequency gain matrix of the VSC part.

Using Table 2.1, Fig. 2.4a, b, the ratio of the output voltages to V_{dc} and θ_{SV} in terms of output voltage frequency and phase for sectors II to VI are tabulated in Table 2.3. For all the sectors, evaluation of the average value of the synthesized output voltages results in the expression described by (2.13).

Table 2.3 Ratio of line voltages to V_{dc}, θ_{SV} for sectors II to VI

	v_{AB}/V_{dc}	v_{BC}/V_{dc}	v_{CA}/V_{dc}	θ_{SV}
Sector II	d_α	d_β	$-d_\alpha - d_\beta$	$\omega_o t + \varphi_0$
Sector III	$-d_\beta$	$d_\alpha + d_\beta$	$-d_\alpha$	$\omega_o t + \varphi_0 - 60°$
Sector IV	$-d_\alpha - d_\beta$	d_α	d_β	$\omega_o t + \varphi_0 - 120°$
Sector V	$-d_\alpha$	$-d_\beta$	$d_\alpha + d_\beta$	$\omega_o t + \varphi_0 - 180°$
Sector VI	d_β	$-d_\alpha - d_\beta$	d_α	$\omega_o t + \varphi_0 - 240°$

Table 2.4 Switching state vectors for CSC

+	–	$\bar{\mathbf{I}}_{\mathbf{in}}$	State vectors
a	c	$\sqrt{2}I_{dc}\angle 30°$	$\mathbf{I}_1$
b	c	$\sqrt{2}I_{dc}\angle 90°$	$\mathbf{I}_2$
b	a	$\sqrt{2}I_{dc}\angle 150°$	$\mathbf{I}_3$
c	a	$\sqrt{2}I_{dc}\angle 210°$	$\mathbf{I}_4$
c	b	$\sqrt{2}I_{dc}\angle 270°$	$\mathbf{I}_5$
a	b	$\sqrt{2}I_{dc}\angle 330°$	$\mathbf{I}_6$
a	a	0	I_0
b	b		
c	c		

The low frequency gain matrix of the CSC part is derived in a similar manner. If φ_i is the desired input displacement angle, the desired input currents are then represented as

$$\mathbf{i}_{\mathbf{in}}^{*\,(3)} = \begin{bmatrix} i_a^* \\ i_b^* \\ i_c^* \end{bmatrix} = \hat{I}_{in} \begin{bmatrix} \cos(\omega_i t - \varphi_c - \varphi_i) \\ \cos(\omega_i t - \varphi_c - \varphi_i - 120°) \\ \cos(\omega_i t - \varphi_c - \varphi_i + 120°) \end{bmatrix}, \tag{2.14}$$

and therefore the target input current rotating vector is

$$\bar{\mathbf{I}}_{\mathbf{in}}^* = \sqrt{\frac{2}{3}}\,(i_a^* + i_b^*\, e^{j2\pi/3} + i_c^*\, e^{-j2\pi/3}) = \sqrt{\frac{3}{2}}\,\hat{I}_{in}\, e^{j(\omega_i t - \varphi_c - \varphi_i)}. \tag{2.15}$$

The valid switching states along with the SVs are tabulated in Table 2.4. Figure 2.5 shows the SVs of the CSC construct.

The duty cycles for the switches of the fictitious CSC part are determined in the same manner as their VSC counterparts. They are

$$d_\mu = \frac{\hat{I}_{in}}{I_{dc}} \sin(60° - \theta_{SI}) = m_i \sin(60° - \theta_{SI}) \;\; \& \;\; d_\gamma = \frac{\hat{I}_{in}}{I_{dc}} \sin(\theta_{SI}) = m_i \sin(\theta_{SI}). \tag{2.16}$$

Table 2.5 Input currents when $\bar{\mathbf{I}}_{\mathbf{in}}^{*}$ is in sector I

	+	−		i_a	i_b	i_c
During $d_6\,(d_\mu)$	a	b	⟶	I_{dc}	$-I_{dc}$	0
During $d_1\,(d_\gamma)$	a	c	⟶	I_{dc}	0	$-I_{dc}$

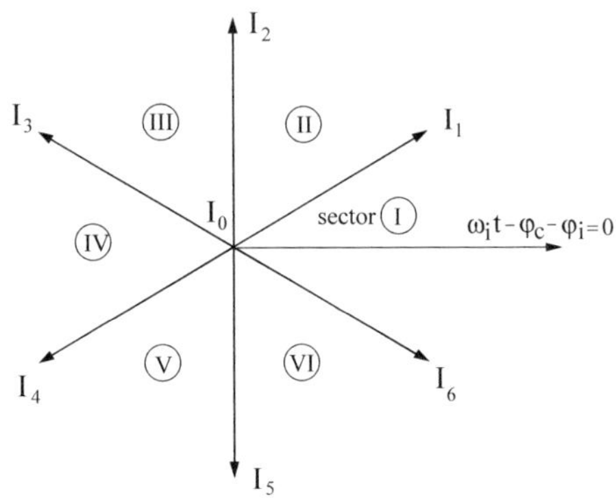

Fig. 2.5 Input current vectors

Here, θ_{SI}, μ, γ are analogous to θ_{SV}, α, β in Fig. 2.4b. When $\bar{\mathbf{I}}_{\mathbf{in}}^{*}$ is in sector I, where $\mathbf{I}_\mu = \mathbf{I}_6$ and $\mathbf{I}_\gamma = \mathbf{I}_1$, the input currents are shown in Table 2.5.

Additionally, for sector I, θ_{SI} can be expressed as

$$\theta_{SI} = \omega_i t - \varphi_c - \varphi_i + 30^\circ. \tag{2.17}$$

Using (2.16), (2.17) and Table 2.5, the synthesized input phase currents are

$$\mathbf{i_{in}}^{(3)} = \begin{bmatrix} i_a \\ i_b \\ i_c \end{bmatrix} = \begin{bmatrix} d_\mu + d_\gamma \\ -d_\mu \\ -d_\gamma \end{bmatrix} I_{dc} = m_i \underbrace{\begin{bmatrix} \cos(\omega_i t - \varphi_c - \varphi_i) \\ \cos(\omega_i t - \varphi_c - \varphi_i - 120^\circ) \\ \cos(\omega_i t - \varphi_c - \varphi_i + 120^\circ) \end{bmatrix}}_{\mathbf{m_i}} I_{dc}$$
$$= \mathbf{m_i} I_{dc}, \tag{2.18}$$

where $\mathbf{m_i}$ represents the low frequency gain matrix of the CSC part. The ratio of the input current to I_{dc} and θ_{SI} in terms of input current frequency and phase for sectors II–VI are tabulated in Table 2.6.

From the discussion based on separate CSC and VSC structure, it seems that both $\mathbf{v_{oL}^{*}}^{(3)}$ and $\mathbf{i_{in}^{*}}^{(3)}$ can be synthesized independently. This would imply that the duration of active vectors in one converter, i.e. those corresponding to active power flow, can be independent of the active power flow duration in the second converter. However this is not possible. As an example, a zero vector for VSC means all 3 output phases are connected to the same DC link terminal—resulting in zero I_{dc}. Hence, a nonzero

Table 2.6 Ratio of input currents to I_{dc}, θ_{SI} for sectors II–VI

	i_a/I_{dc}	i_b/I_{dc}	i_c/I_{dc}	θ_{SI}
Sector II	d_μ	d_γ	$-d_\mu - d_\gamma$	$\omega_i t - \varphi_c - \varphi_i - 30^\circ$
Sector III	$-d_\gamma$	$d_\mu + d_\gamma$	$-d_\mu$	$\omega_i t - \varphi_c - \varphi_i - 90^\circ$
Sector IV	$-d_\mu - d_\gamma$	d_μ	d_γ	$\omega_i t - \varphi_c - \varphi_i - 150^\circ$
Sector V	$-d_\mu$	$-d_\gamma$	$d_\mu + d_\gamma$	$\omega_i t - \varphi_c - \varphi_i - 210^\circ$
Sector VI	d_γ	$-d_\mu - d_\gamma$	d_μ	$\omega_i t - \varphi_c - \varphi_i - 270^\circ$

$\bar{\mathbf{I}}^*_{\mathbf{in}}$ cannot be synthesized during this period. Therefore, at a given instant of time, it is not possible to have an active vector in one converter and a zero vector in the other one. This is where the second stage of duty cycle computation begins as described in the following.

Absence of any intermediate storage element in the power stage of MC implies that instantaneous active power at the input and output side must be same. This power balance necessarily requires that the active vector states of VSC and CSC must be chosen at the same instant and applied for exactly the same duration. This requires combining the gain matrices of the VSC and CSC parts. If P_{in} be the throughput power, then using (2.18),

$$P_{in} = \{\mathbf{i_{in}}^{(3)}\}^T \mathbf{v_{cf}}^{(3)} = \{\mathbf{m_i}\}^T I_{dc} \mathbf{v_{cf}}^{(3)} = \{\mathbf{m_i}\}^T \frac{P_{in}}{V_{dc}} \mathbf{v_{cf}}^{(3)} \implies V_{dc} = \{\mathbf{m_i}\}^T \mathbf{v_{cf}}^{(3)}. \tag{2.19}$$

Subsequently, substituting V_{dc} in (2.13) by using (2.19), the voltage gain relation is obtained as

$$\mathbf{v_{oL}}^{(3)} = \mathbf{m_v}\{\mathbf{m_i}\}^T \mathbf{v_{cf}}^{(3)}. \tag{2.20}$$

Hence $\mathbf{m_v}\{\mathbf{m_i}\}^T$ is the input phase to output line voltage low frequency gain matrix. Using (2.13) and (2.18), this gain is evaluated as

$$\mathbf{m_v}\{\mathbf{m_i}\}^T = \underbrace{m_v m_i}_{m} \begin{bmatrix} \cos(\omega_o t + \varphi_o + 30^\circ) \\ \cos(\omega_o t + \varphi_o + 30^\circ - 120^\circ) \\ \cos(\omega_o t + \varphi_o + 30^\circ + 120^\circ) \end{bmatrix} \begin{bmatrix} \cos(\omega_i t - \varphi_c - \varphi_i) \\ \cos(\omega_i t - \varphi_c - \varphi_i - 120^\circ) \\ \cos(\omega_i t - \varphi_c - \varphi_i + 120^\circ) \end{bmatrix}^T, \tag{2.21}$$

where m is commonly referred to as modulation index. The output voltage referred to output neutral n, can thus be represented as

$$\mathbf{v_{on}}^{(3)} = \mathbf{m}^{(3)} \mathbf{v_{cf}}^{(3)}, \tag{2.22}$$

where, the input-output phase voltage gain matrix $\mathbf{m}^{(3)}$ is obtained from (2.21) as

$$\mathbf{m}^{(3)} = (m/\sqrt{3}) \begin{bmatrix} \cos(\omega_o t + \varphi_o) \\ \cos(\omega_o t + \varphi_o - 120°) \\ \cos(\omega_o t + \varphi_o + 120°) \end{bmatrix} \begin{bmatrix} \cos(\omega_i t - \varphi_c - \varphi_i) \\ \cos(\omega_i t - \varphi_c - \varphi_i - 120°) \\ \cos(\omega_i t - \varphi_c - \varphi_i + 120°) \end{bmatrix}^T . \tag{2.23}$$

From power balance and gain relationship of (2.22),

$$\mathbf{i_{in}}^{(3)} = \left(\mathbf{m}^{(3)}\right)^T \mathbf{i_o}^{(3)}. \tag{2.24}$$

Instantaneous input-output power balance also implies that

$$\frac{3}{2}\hat{V}_{cf}\hat{I}_{in}\cos\varphi_i = \frac{3}{2}\hat{V}_o\hat{I}_o\cos\varphi_{oL}, \tag{2.25}$$

where, φ_{oL} is the output displacement angle and $\hat{I}_o$ is the amplitude of the output current (i_o). So for a given $\mathbf{v_{on}^*}^{(3)}$, load and $\mathbf{v_{cf}}^{(3)}$, it is not possible to independently control φ_i and $\hat{I}_{in}$. Using (2.25) along with m_v and m_i defined in (2.10) and (2.16) respectively, the modulation index m is evaluated as

$$m = m_v m_i = \left(\frac{\sqrt{3}\,\hat{V}_o}{V_{dc}}\right)\left(\frac{\hat{I}_{in}}{I_{dc}}\right) = \frac{\sqrt{3}\,\hat{V}_o\hat{I}_{in}}{P_{in}} = \frac{\hat{V}_o}{\frac{\sqrt{3}}{2}\hat{V}_{cf}\cos(\varphi_i)}. \tag{2.26}$$

Using (2.25) and (2.26), the magnitude of the input output variables are related as

$$\hat{V}_o = \frac{\sqrt{3}}{2} m\,\hat{V}_{cf}\cos(\varphi_i) \quad \text{and} \quad \hat{I}_{in} = \frac{\sqrt{3}}{2} m\,\hat{I}_o\cos(\varphi_{oL}). \tag{2.27}$$

Now, when both $\bar{\mathbf{V}}_{\mathbf{oL}}^*$ and $\bar{\mathbf{I}}_{\mathbf{in}}^*$ are in sector I of their corresponding hexagons, using (2.11) and (2.18), the input phase to output line voltage low frequency gain matrix is obtained as

$$\mathbf{m_v}\{\mathbf{m_i}\}^T = \begin{bmatrix} d_\alpha + d_\beta \\ -d_\alpha \\ -d_\beta \end{bmatrix} \begin{bmatrix} d_\mu + d_\gamma \\ -d_\mu \\ -d_\gamma \end{bmatrix}^T . \tag{2.28}$$

Therefore the average output line voltages over a switching cycle T_s are

$$\begin{bmatrix} v_{AB} \\ v_{BC} \\ v_{CA} \end{bmatrix} = \begin{bmatrix} d_\alpha + d_\beta \\ -d_\alpha \\ -d_\beta \end{bmatrix} \begin{bmatrix} d_\mu + d_\gamma \\ -d_\mu \\ -d_\gamma \end{bmatrix}^T \begin{bmatrix} v_a \\ v_b \\ v_c \end{bmatrix}$$

$$= \begin{bmatrix} d_\mu(d_\alpha + d_\beta)(v_a - v_b) - d_\gamma(d_\alpha + d_\beta)(v_c - v_a) \\ -d_\alpha d_\mu(v_a - v_b) + d_\alpha d_\gamma(v_c - v_a) \\ -d_\beta d_\mu(v_a - v_b) + d_\beta d_\gamma(v_c - v_a) \end{bmatrix}$$

$$= \begin{bmatrix} d_\mu(d_\alpha + d_\beta) \\ -d_\alpha d_\mu \\ -d_\beta d_\mu \end{bmatrix} v_{ab} + \begin{bmatrix} -d_\gamma(d_\alpha + d_\beta) \\ d_\alpha d_\gamma \\ d_\beta d_\gamma \end{bmatrix} v_{ca}. \tag{2.29}$$

Thus there are four combinations of duty cycles $d_\alpha d_\mu$, $d_\alpha d_\gamma$, $d_\beta d_\mu$ and $d_\beta d_\gamma$ corresponding to the active states $V_\alpha I_\mu$, $V_\alpha I_\gamma$, $V_\beta I_\mu$ and $V_\beta I_\gamma$. Using (2.10), (2.16) and (2.26), the duty cycle for the active state $V_\alpha I_\mu$ is

$$\begin{aligned} d_\alpha d_\mu &= \frac{\sqrt{3}\,\hat{V}_o}{V_{dc}} \frac{\hat{I}_{in}}{I_{dc}} \sin(60^\circ - \theta_{SV}) \sin(60^\circ - \theta_{SI}) \\ &= m \sin(60^\circ - \theta_{SV}) \sin(60^\circ - \theta_{SI}). \end{aligned} \tag{2.30}$$

Similarly, the following are derived

$$\begin{aligned} d_\alpha d_\gamma &= m \sin(60^\circ - \theta_{SV}) \sin(\theta_{SI}), \\ d_\beta d_\mu &= m \sin(\theta_{SV}) \sin(60^\circ - \theta_{SI}), \\ d_\beta d_\gamma &= m \sin(\theta_{SV}) \sin(\theta_{SI}). \end{aligned} \tag{2.31}$$

The input-output connections over a switching period in the decoupled construct and the corresponding connections in the actual matrix topology are shown in Table 2.7.

Table 2.7 Mapping from CSC VSC construct to 9 Qsw MC

	CSC VSC construct					3 Ph MC		
ON time	A	B	C	+	–	A	B	C
$d_\alpha d_\mu T_S$	+	–	+	a	b	a	b	a
$d_\beta d_\mu T_S$	+	–	–	a	b	a	b	b
$d_\alpha d_\gamma T_S$	+	–	+	a	c	a	c	a
$d_\beta d_\gamma T_S$	+	–	–	a	c	a	c	c

2.1.1.1 Transformation of 3 Phase Variables in *abc* Domain to *dq* Domain

In this analysis the modelling of the system has been carried out in a synchronously rotating (dq) domain. The advantage of working in the dq domain are that the fundamental frequency components of all variables which were varying sinusoidally in abc domain are transformed to time invariant quantities in steady state. This greatly simplifies the task of analysis and controller design.

The 3 Ph variables of (2.22) in abc domain are transformed to a synchronously rotating (dq) frame using the transfer matrix

$$\mathbf{T} = \sqrt{\frac{2}{3}} \begin{bmatrix} \cos(\omega_T t) & \cos(\omega_T t - 120^\circ) & \cos(\omega_T t + 120^\circ) \\ \sin(\omega_T t) & \sin(\omega_T t - 120^\circ) & \sin(\omega_T t + 120^\circ) \\ 1/\sqrt{2} & 1/\sqrt{2} & 1/\sqrt{2} \end{bmatrix}, \tag{2.32}$$

in the following manner

$$\mathbf{v_{on}} = \begin{bmatrix} v_{ond} \\ v_{onq} \\ v_{on0} \end{bmatrix} = \mathbf{T}\mathbf{v_{on}}^{(3)} = \left(\mathbf{T}\mathbf{m}^{(3)}\mathbf{T}^{-1}\right)\mathbf{T}\mathbf{v_{cf}}^{(3)} = \mathbf{m} \begin{bmatrix} v_{cfd} \\ v_{cfq} \\ v_{cf0} \end{bmatrix} = \mathbf{m}\,\mathbf{v_{cf}}. \tag{2.33}$$

Here, $\omega_T t$ is estimated by a PLL locked to the phase—a of the point of common coupling (PCC). For synchronous applications, $\omega_T = \omega_o = \omega_i$. Since balanced 3-phase system has been considered, the zero sequence components can be neglected. The output-input voltage relationship in (2.33) gets modified to

$$\mathbf{v_{on}} = \begin{bmatrix} v_{ond} \\ v_{onq} \end{bmatrix} = \mathbf{m}\,\mathbf{v_{cf}} = \mathbf{m} \begin{bmatrix} v_{cfd} \\ v_{cfq} \end{bmatrix}, \tag{2.34}$$

where, the gain matrix $\mathbf{m}$ is obtained using (2.22), (2.23) and (2.32), as

$$\mathbf{m} = \frac{\sqrt{3}}{2} m \begin{bmatrix} \cos\varphi_o \\ \sin\varphi_o \end{bmatrix} \begin{bmatrix} \cos\theta_i & \sin\theta_i \end{bmatrix} = \begin{bmatrix} m_d \\ m_q \end{bmatrix} \begin{bmatrix} \cos\theta_i & \sin\theta_i \end{bmatrix}. \tag{2.35}$$

Here,

$$\theta_i = \varphi_i + \varphi_c. \tag{2.36}$$

m_d and m_q allow complete control on the voltage gain of MC and θ_i on the input displacement angle.

The input currents are obtained as

$$\mathbf{i_{in}} = \mathbf{m}^T \mathbf{i_o}. \tag{2.37}$$

Fig. 2.6 Low frequency gain of MC

This concludes the discussion on the low frequency gain of MC. The 'black box' model of the low frequency gain of MC is shown in Fig. 2.6. Dynamic modelling is discussed in the next section.

2.2 Linearized Model

Figure 2.7 shows the single phase equivalent of the overall system where the source impedance has been modelled as an inductance L_s. **P** represents the PCC. An output ripple filter is also necessary to attenuate the switching frequency components in the voltage waveforms.

Referring to Fig. 2.7, dynamic equations for i_o and v_{cf} in abc domain are,

$$\frac{d}{dt}\mathbf{i_o}^{(3)} = \frac{1}{L_o}\left(\mathbf{v_{on}}^{(3)} - \mathbf{v_{co}}^{(3)}\right) \quad \text{and} \quad \frac{d}{dt}\mathbf{v_{cf}}^{(3)} = \frac{1}{C_f}\left(\mathbf{i_s}^{(3)} - \mathbf{i_{in}}^{(3)}\right). \tag{2.38}$$

Substituting $\mathbf{v_{on}}^{(3)}$ and $\mathbf{i_{in}}^{(3)}$ in (2.38) by applying the gain equations (2.22) and (2.24), leads to

$$\frac{d}{dt}\mathbf{i_o}^{(3)} = \frac{1}{L_o}\left(\mathbf{m}^{(3)}\mathbf{v_{cf}}^{(3)} - \mathbf{v_{co}}^{(3)}\right) \quad \text{and} \quad \frac{d}{dt}\mathbf{v_{cf}}^{(3)} = \frac{1}{C_f}\left(\mathbf{i_s}^{(3)} - \left(\mathbf{m}^{(3)}\right)^T \mathbf{i_o}^{(3)}\right). \tag{2.39}$$

These equations are transformed to dq domain using the transfer matrix **T** in (2.32). Subsequent omission of the zero sequence components in the transformed equations results in

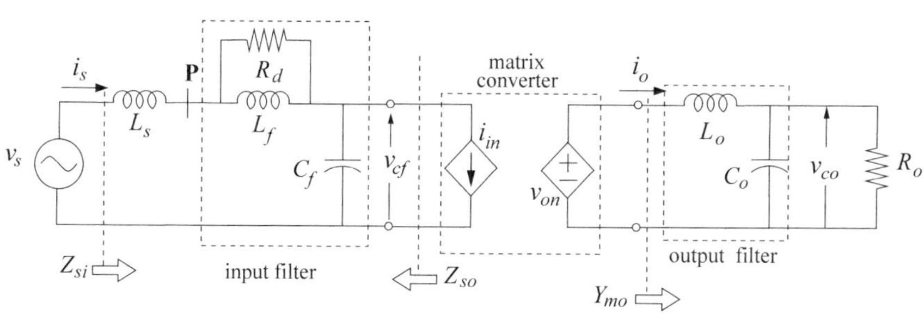

Fig. 2.7 Single phase diagram including source inductance

$$\frac{d}{dt}\mathbf{i}_o = \begin{bmatrix} 0 & -\omega_T \\ \omega_T & 0 \end{bmatrix} \mathbf{i_o} + \frac{1}{L_o}\left(\mathbf{m}\mathbf{v_{cf}} - \mathbf{v_{co}}\right), \tag{2.40}$$

$$\frac{d}{dt}\mathbf{v}_{cf} = \begin{bmatrix} 0 & -\omega_T \\ \omega_T & 0 \end{bmatrix} \mathbf{v_{cf}} + \frac{1}{C_f}\left(\mathbf{i_s} - \mathbf{m}^T\mathbf{i_o}\right). \tag{2.41}$$

Equations (2.40) and (2.41) are nonlinear as product of states ($\mathbf{v_{cf}}$, $\mathbf{i_o}$) and inputs ($\mathbf{m}$) are involved. These are linearized around an equilibrium point (I_{sd}, I_{sq}, I_{Lfd}, I_{Lfq}, V_{cfd}, V_{cfq}, I_{od}, I_{oq}, V_{cod}, V_{coq}, V_{sd} and V_{sq}). Denoting equilibrium value of a variable x as X and a small perturbation in x around X as $\tilde{x}$, the equations corresponding to (2.40) and (2.41) in the linearized model are

$$\frac{d}{dt}\tilde{\mathbf{i}}_o = \begin{bmatrix} 0 & -\omega_T \\ \omega_T & 0 \end{bmatrix} \tilde{\mathbf{i}}_o + \frac{1}{L_o}\left(\tilde{\mathbf{m}}\mathbf{V_{cf}} + \mathbf{M}\,\tilde{\mathbf{v}}_{\mathbf{cf}} - \tilde{\mathbf{v}}_{\mathbf{co}}\right), \tag{2.42}$$

$$\frac{d}{dt}\tilde{\mathbf{v}}_{\mathbf{cf}} = \begin{bmatrix} 0 & -\omega_T \\ \omega_T & 0 \end{bmatrix} \tilde{\mathbf{v}}_{\mathbf{cf}} + \frac{1}{C_f}\left(\tilde{\mathbf{i}}_s - \tilde{\mathbf{m}}^T\mathbf{I_o} - \mathbf{M}^T\,\tilde{\mathbf{i}}_{\mathbf{o}}\right). \tag{2.43}$$

$\tilde{\mathbf{m}}$ and $\mathbf{M}$ are obtained using (2.35) as,

$$\mathbf{m} = \mathbf{M} + \tilde{\mathbf{m}} = \begin{bmatrix} M_d + \tilde{m}_d \\ M_q + \tilde{m}_q \end{bmatrix} \begin{bmatrix} \cos(\theta_i + \tilde{\theta}_i) \\ \sin(\theta_i + \tilde{\theta}_i) \end{bmatrix}^T$$

$$= \underbrace{\begin{bmatrix} M_d \\ M_q \end{bmatrix} \begin{bmatrix} C_\theta & S_\theta \end{bmatrix}}_{\mathbf{M}} + \underbrace{\begin{bmatrix} \tilde{m}_d \\ \tilde{m}_q \end{bmatrix} \begin{bmatrix} C_\theta & S_\theta \end{bmatrix} + \begin{bmatrix} -M_d S_\theta & M_d C_\theta \\ -M_q S_\theta & M_q C_\theta \end{bmatrix} \tilde{\theta}_i}_{\tilde{\mathbf{m}}} \tag{2.44}$$

where

$$\cos\theta_i = C_\theta \quad \text{and} \quad \sin\theta_i = S_\theta. \tag{2.45}$$

So far, the analysis was carried out in time domain to highlight the source of nonlinearity in the dynamic equations and the consequent requirement of linearizing them. Continuing the system description with dynamic equations in time domain invariably leads to state space description and subsequently eigenvalue analysis using computational tools. It is very difficult to correlate the eigenvalues with the passive components, among which the filter elements are actually design parameters. Consequently, an understanding of how the system dynamics are affected by hardware design and different operating points cannot be developed. Hence, this route has been consciously avoided in this book. Therefore, the system modelling from now onwards will be discussed in s domain which enables use of transfer functions.

Before detailing the 3 Ph small signal model, the transfer functions of interest in the per-phase system of Fig. 2.7 are described. The input filter output impedance, Z_{so} is defined as

$$Z_{so}(s) = -\frac{v_{cf}(s)}{i_{in}(s)}\bigg|_{v_s(s)=0} = \frac{s^2L_sL_f + R_d(L_s+L_f)s}{s^3L_sL_fC_f + s^2(L_s+L_f)R_dC_f + sL_f + R_d}. \tag{2.46}$$

Input impedance seen from source, Z_{si} is

$$Z_{si}(s) = \frac{v_s(s)}{i_s(s)}\bigg|_{i_{in}(s)=0} = \frac{s^3L_sL_fC_f + s^2(L_s+L_f)R_dC_f + sL_f + R_d}{(sL_f + R_d)sC_f}, \tag{2.47}$$

and the output admittance seen from the converter output terminals Y_{mo}, is

$$Y_{mo}(s) = \frac{1}{Z_{mo}(s)} = \frac{i_o(s)}{v_{on}(s)} = \frac{1}{R_o}\frac{1 + sR_oC_o}{s^2L_oC_o + s\frac{L_o}{R_o} + 1}. \tag{2.48}$$

The forward gain for the input and output filter are

$$G_{fv}(s) = \frac{v_{cf}(s)}{v_s(s)}\bigg|_{i_{in}(s)=0} = \frac{i_s(s)}{i_{in}(s)}\bigg|_{v_s(s)=0} = \frac{1}{sC_f}\frac{1}{Z_{si}(s)}, \tag{2.49}$$

$$G_{vof}(s) = \frac{v_{co}(s)}{v_{on}(s)} = \frac{1}{s^2L_oC_o + s\frac{L_o}{R_o} + 1}. \tag{2.50}$$

Hence in the per-phase system, $i_s(s)$ can be described as

$$\begin{aligned} i_s(s) &= \frac{i_s(s)}{i_{in}(s)}\bigg|_{v_s=0} i_{in}(s) \quad + \quad \frac{i_s(s)}{v_s(s)}\bigg|_{i_{\mathbf{in}}=0} v_s(s) \\ &= G_{fv}(s)i_{in}(s) \quad + \quad Z_{si}^{-1}(s)v_s(s). \end{aligned} \tag{2.51}$$

Therefore in *abc* domain for a balanced 3 Ph system,

$$\mathbf{i_s}^{(3)}(s) = \mathbf{I}_3 G_{fv}(s)\mathbf{i_{in}}^{(3)}(s) \quad + \quad \mathbf{I}_3 Z_{si}^{-1}(s)\mathbf{v_s}^{(3)}(s), \tag{2.52}$$

where $\mathbf{I}_3$ represents a 3×3 identity matrix.

Whenever any transfer function matrix for a balanced 3 Ph system $\mathbf{I}_3G(s)$ in *abc* domain is transformed to dq domain, it results in a non-diagonal (coupled) transfer matrix $\mathbf{G}(s)$ [41]. The transformed matrix $\mathbf{G}(s)$ is represented in the form

$$\mathbf{G}(s) = \begin{bmatrix} G_{11}(s) & G_{12}(s) \\ -G_{12}(s) & G_{11}(s) \end{bmatrix}, \tag{2.53}$$

where the individual elements are

$$G_{11}(s) = \frac{1}{2}\,\{G(s + j\omega_T) + G(s - j\omega_T)\},$$
$$G_{12}(s) = -\frac{j}{2}\,\{G(s + j\omega_T) - G(s - j\omega_T)\}. \tag{2.54}$$

For example, $\mathbf{I_3}G_{fv}(s)$ is transformed to $\mathbf{G_{fv}}(s)$ as

$$\mathbf{G_{fv}}(s) = \begin{bmatrix} G_{fv11}(s) & G_{fv12}(s) \\ -G_{fv12}(s) & G_{fv11}(s) \end{bmatrix}, \quad \text{where,}$$
$$G_{fv11}(s) = \frac{1}{2}\,\left\{G_{fv}(s + j\omega_T) + G_{fv}(s - j\omega_T)\right\} \ \ \&$$
$$G_{fv12}(s) = -\frac{j}{2}\,\left\{G_{fv}(s + j\omega_T) - G_{fv}(s - j\omega_T)\right\}. \tag{2.55}$$

$\mathbf{Z_{si}^{-1}}(s)$ can be similarly derived from its per phase counterpart $Z_{si}^{-1}(s)$. Hence (2.52) can be described in dq domain as

$$\mathbf{i_s}(s) = \mathbf{G_{fv}}(s)\mathbf{i_{in}}(s) \quad + \quad \mathbf{Z_{si}^{-1}}(s)\mathbf{v_s}(s). \tag{2.56}$$

From (2.56), the perturbed variable $\tilde{\mathbf{i}}_\mathbf{s}(s)$ in the linear model can be described as

$$\tilde{\mathbf{i}}_\mathbf{s}(s) = \mathbf{G_{fv}}(s)\,\tilde{\mathbf{i}}_\mathbf{in}(s) + \mathbf{Z_{si}^{-1}}(s)\tilde{\mathbf{v}}_\mathbf{s}(s). \tag{2.57}$$

Using the transfer matrix notation of (2.53), (2.57) can be expanded as

$$\begin{bmatrix} \tilde{i}_{sd}(s) \\ \tilde{i}_{sq}(s) \end{bmatrix} = \begin{bmatrix} G_{fv11}(s) & G_{fv12}(s) \\ -G_{fv12}(s) & G_{fv11}(s) \end{bmatrix} \begin{bmatrix} \tilde{i}_{ind}(s) \\ \tilde{i}_{inq}(s) \end{bmatrix} + \begin{bmatrix} Z_{si11}^{-1}(s) & Z_{si12}^{-1}(s) \\ -Z_{si12}^{-1}(s) & Z_{si11}^{-1}(s) \end{bmatrix} \begin{bmatrix} \tilde{v}_{sd}(s) \\ \tilde{v}_{sq}(s) \end{bmatrix}. \tag{2.58}$$

Similarly, other perturbed variables in dq domain can be described using the per-phase equations (2.46)–(2.50), as

$$\tilde{\mathbf{v}}_\mathbf{cf}(s) = \mathbf{G_{fv}}(s)\tilde{\mathbf{v}}_\mathbf{s}(s) - \mathbf{Z_{so}}(s)\,\tilde{\mathbf{i}}_\mathbf{in}(s), \tag{2.59}$$
$$\tilde{\mathbf{i}}_\mathbf{o}(s) = \mathbf{Y_{mo}}(s)\,\tilde{\mathbf{v}}_\mathbf{on}(s) \ = \ \mathbf{Z_{mo}^{-1}}(s)\,\tilde{\mathbf{v}}_\mathbf{on}(s), \tag{2.60}$$
$$\tilde{\mathbf{v}}_\mathbf{co}(s) = \mathbf{G_{vof}}(s)\,\tilde{\mathbf{v}}_\mathbf{on}(s). \tag{2.61}$$

Using (2.34), (2.37) and (2.44), the gain equations in the linearized model are obtained as

$$\tilde{\mathbf{v}}_\mathbf{on}(s) = \tilde{\mathbf{m}}(s)\mathbf{V_{cf}} + \mathbf{M}\,\tilde{\mathbf{v}}_\mathbf{cf}(s), \tag{2.62}$$
$$\tilde{\mathbf{i}}_\mathbf{in}(s) = (\tilde{\mathbf{m}}(s))^T\,\mathbf{I_o} + \mathbf{M}^T\,\tilde{\mathbf{i}}_\mathbf{o}(s). \tag{2.63}$$

The 3 Ph small signal model in dq domain is completely described by (2.57)–(2.63), the equivalent circuit of which is shown in Fig. 2.8. Figure 2.8a, b represent

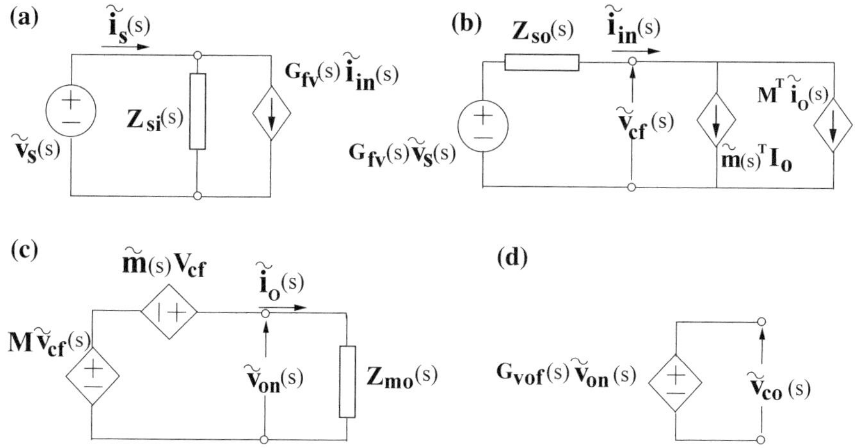

Fig. 2.8 Equivalent circuit of the linearized system. **a**, **b** Input, **c**, **d** Output side

the input side described by (2.57), (2.59) and (2.63). The output side described by (2.60), (2.61) and (2.62) is represented by Fig. 2.8c, d.

The output variables in this model are the voltages $\tilde{\mathbf{v}}_{\mathbf{co}}(s)$ across the output filter capacitors and the q-axis (reactive) component of source current, $\tilde{i}_{sq}(s)$. From (2.44), the control inputs are $\tilde{m}_d(s)$, $\tilde{m}_q(s)$ and $\tilde{\theta}_i(s)$. In the present analysis, the output filter voltages are the only variables which are being controlled. Thus, θ_i is assumed to be set at an arbitrary constant value θ_I, leaving $\left[\tilde{m}_d(s)\,\tilde{m}_q(s)\right]^T$ as the only control inputs while $\tilde{\mathbf{v}}_{\mathbf{co}}(s)(=\left[\tilde{v}_{cod}(s)\,\tilde{v}_{coq}(s)\right]^T)$ are the output variables of the dynamic model. The steps for deriving the input to output transfer matrix are described subsequently, where, the disturbance inputs $\tilde{\mathbf{v}}_{\mathbf{s}}(s)$ are not considered.

Substituting $\tilde{\mathbf{i}}_{\mathbf{in}}(s)$ from (2.63) in (2.59), and dropping the Laplace operator in the notation yields

$$\tilde{\mathbf{v}}_{\mathbf{cf}} = -\mathbf{Z}_{\mathbf{so}}\tilde{\mathbf{m}}^T\mathbf{I}_{\mathbf{o}} - \mathbf{Z}_{\mathbf{so}}\mathbf{M}^T\,\tilde{\mathbf{i}}_{\mathbf{o}}. \tag{2.64}$$

Subsequent substitution of $\tilde{\mathbf{i}}_{\mathbf{o}}$ from (2.60) in (2.64) results in

$$\tilde{\mathbf{v}}_{\mathbf{cf}} = -\mathbf{Z}_{\mathbf{so}}\tilde{\mathbf{m}}^T\mathbf{I}_{\mathbf{o}} - \mathbf{Z}_{\mathbf{so}}\mathbf{M}^T\,\mathbf{Y}_{\mathbf{mo}}\,\tilde{\mathbf{v}}_{\mathbf{on}}. \tag{2.65}$$

Thereafter substituting $\tilde{\mathbf{v}}_{\mathbf{cf}}$ from (2.65) in (2.62) yields

$$\tilde{\mathbf{v}}_{\mathbf{on}} = \tilde{\mathbf{m}}\mathbf{V}_{\mathbf{cf}} - \mathbf{M}\,\mathbf{Z}_{\mathbf{so}}\tilde{\mathbf{m}}^T\mathbf{I}_{\mathbf{o}} - \mathbf{M}\,\mathbf{Z}_{\mathbf{so}}\mathbf{M}^T\,\mathbf{Y}_{\mathbf{mo}}\,\tilde{\mathbf{v}}_{\mathbf{on}}$$

$$\Longrightarrow \left(\mathbf{I}_2 + \mathbf{M}\mathbf{Z}_{\mathbf{so}}\mathbf{M}^T\mathbf{Y}_{\mathbf{mo}}\right)\tilde{\mathbf{v}}_{\mathbf{on}} = \left(\tilde{\mathbf{m}}\mathbf{V}_{\mathbf{cf}} - \mathbf{M}\mathbf{Z}_{\mathbf{so}}\tilde{\mathbf{m}}^T\mathbf{I}_{\mathbf{o}}\right). \tag{2.66}$$

where $\mathbf{I_2}$ represents a 2×2 identity matrix. Denoting $\mathbf{I_2} + \mathbf{M}\mathbf{Z_{so}}\mathbf{M}^T\mathbf{Y_{mo}}$ as

$$\mathbf{G_I} = \mathbf{I_2} + \mathbf{M}\mathbf{Z_{so}}\mathbf{M}^T\mathbf{Y_{mo}}, \tag{2.67}$$

$\tilde{\mathbf{v}}_{\mathbf{on}}$ in (2.66) can be expressed as

$$\tilde{\mathbf{v}}_{\mathbf{on}} = \mathbf{G_I}^{-1}\left(\tilde{\mathbf{m}}\mathbf{V_{cf}} - \mathbf{M}\mathbf{Z_{so}}\tilde{\mathbf{m}}^T\mathbf{I_o}\right). \tag{2.68}$$

Expanded form of $\mathbf{G_I}^{-1}$ is derived in (A.3) of Appendix A.1. It has also been shown in Appendix A.2 that $\tilde{\mathbf{m}}\mathbf{V_{cf}} - \mathbf{M}\mathbf{Z_{so}}\tilde{\mathbf{m}}^T\mathbf{I_o}$ can be simplified to

$$\left(\tilde{\mathbf{m}}\mathbf{V_{cf}} - \mathbf{M}\mathbf{Z_{so}}\tilde{\mathbf{m}}^T\mathbf{I_o}\right) = \mathbf{G_a}(s)\begin{bmatrix}\tilde{m}_d \\ \tilde{m}_q\end{bmatrix}, \tag{2.69}$$

where

$$\mathbf{G_a}(s) = \begin{bmatrix} a_V - M_d Z_{so11} I_{od} & -M_d Z_{so11} I_{oq} \\ -M_q Z_{so11} I_{od} & a_V - M_q Z_{so11} I_{oq} \end{bmatrix} \tag{2.70}$$

and

$$a_V = \sqrt{\frac{3}{2}}\,\hat{V}_{cf}\cos\varphi_i. \tag{2.71}$$

Therefore using (2.68) and (2.69),

$$\tilde{\mathbf{v}}_{\mathbf{on}} = \mathbf{G_I}^{-1}\mathbf{G_a}\begin{bmatrix}\tilde{m}_d \\ \tilde{m}_q\end{bmatrix}. \tag{2.72}$$

Finally, substituting $\tilde{\mathbf{v}}_{\mathbf{on}}$ from (2.72) in (2.61), the input to output equations after re-introducing the Laplace operator are

$$\tilde{\mathbf{v}}_{\mathbf{co}}(s) = \mathbf{G_{vof}}(s)\tilde{\mathbf{v}}_{\mathbf{on}}(s) = \mathbf{G_{vof}}(s)\mathbf{G_I}^{-1}(s)\mathbf{G_a}(s)\begin{bmatrix}\tilde{m}_d(s) \\ \tilde{m}_q(s)\end{bmatrix} = \mathbf{G_c}(s)\begin{bmatrix}\tilde{m}_d(s) \\ \tilde{m}_q(s)\end{bmatrix}. \tag{2.73}$$

The control plant transfer function matrix $\mathbf{G_c}(s)$ is a function of both input and output filter parameters as well as the operating points. However it is not apparent how these factors affect the poles and zeros of $\mathbf{G_c}(s)$. From the perspective of controller design the question that immediately arises is whether the plant is stable under all operating conditions. Another related concern is the possibility of non-minimum phase zeros appearing in the plant. These are discussed in the following section.

2.3 Composition of $\mathbf{G_c}(s)$

The first objective of the analysis is to examine whether the plant is stable at all operating points i.e. to detect the presence of any right half poles (RHPs) in $\mathbf{G_c}(s)$. This is accomplished by looking into the 3 component matrices of $\mathbf{G_c}(s)$. Also, another objective is to find out whether the plant contains any right half zeros (RHZs).

It has been shown in Appendix A.3, that all the component matrices of $\mathbf{G_c}(s)$ belong to a special category, where all four elements in each of the component matrices have a common denominator polynomial. Therefore all entries in $\mathbf{G_a}(s)$ have a common denominator polynomial and the same holds true for both $\mathbf{G_I^{-1}}(s)$ and $\mathbf{G_{vof}}(s)$. Analysis of poles and zeros for this specific sub class of common denominator transfer function matrices (CdTM) are described in the following section.

2.3.1 Poles and Zeros of CdTM

A 2×2 transfer matrix $\mathbf{G}(s)$ with normal rank 2 is considered where

$$\mathbf{G}(s) = \begin{bmatrix} G_{11}(s) \; G_{12}(s) \\ G_{21}(s) \; G_{22}(s) \end{bmatrix} = \frac{1}{d(s)} \begin{bmatrix} N_{11}(s) \; N_{12}(s) \\ N_{21}(s) \; N_{22}(s) \end{bmatrix}. \tag{2.74}$$

$d(s)$ is the common denominator polynomial of each elements and $N_{11}(s)$, $N_{12}(s)$, $N_{21}(s)$ and $N_{22}(s)$ are the numerator polynomials. The common denominator $d(s)$ and the matrix composed of numerator polynomials will be denoted here as

$$C_d(\mathbf{G}) = d(s) \quad \text{and} \quad \mathbf{N}(\mathbf{G}) = \begin{bmatrix} N_{11}(s) \; N_{12}(s) \\ N_{21}(s) \; N_{22}(s) \end{bmatrix}.s \tag{2.75}$$

Identifying the poles and zeros of the 2×2 matrix $\mathbf{G}(s)$, requires deriving a diagonal matrix $\mathbf{L}(s)$ from $\mathbf{G}(s)$ as [42, 43],

$$\mathbf{L}(s) = \mathbf{U1}(s)\mathbf{G}(s)\mathbf{U2}(s). \tag{2.76}$$

Here $\mathbf{U1}(s)$ and $\mathbf{U2}(s)$ are the pre and post multiplication matrices used to transform $\mathbf{G}(s)$ to the diagonal matrix $\mathbf{L}(s)$. Therefore determinant of both $\mathbf{U1}(s)$ and $\mathbf{U2}(s)$ are constant. $\mathbf{L}(s)$ is obtained as

$$\mathbf{L}(s) = \begin{bmatrix} \frac{a_1(s)}{b_1(s)} & 0 \\ 0 & \frac{a_2(s)}{b_2(s)} \end{bmatrix}, \tag{2.77}$$

where $a_1(s)$, $b_2(s)$ are factors of $a_2(s)$ and $b_1(s)$ respectively. Then the poles and zeros of $\mathbf{G}(s)$ are obtained as the roots of the following equations

$$p(s) = b_1(s)b_2(s) = 0 \quad \& \quad z(s) = a_1(s)a_2(s) = 0. \tag{2.78}$$

So at any frequency, if control on any single output is lost, it indicates that a zero of the multivariable system resides at that particular frequency.

An alternative procedure for finding the poles and zeros is [42, 44]

- pole polynomial is the least common denominator of all non-zero minors of all orders of $\mathbf{G}(s)$.
- zero polynomial is the greatest common divisor of the minor of order 2 i.e. the determinant of $\mathbf{G}(s)$, where the determinant has been adjusted to have the pole polynomial as the denominator.

Without using computational softwares, finding all the pole and zeros, adhering to the described procedures, is a difficult exercise if the order of the polynomials are high. This difficulty becomes greater if the transfer function matrix is composed of several component matrices as in $\mathbf{G_c}(s)$. For the purpose of controller design, it is important to correctly identify non-minimum phase poles and zeros, if there are any. As would be seen in a later chapter, this knowledge is sufficient for designing a controller.

In (2.77), $\mathbf{L}(s)$ is obtained by applying standard matrix row/column operations on $\mathbf{G}(s)$ and therefore determinant of each is a constant multiple of other. Therefore the determinant of $\mathbf{G}(s)$ will be probed, where,

$$det(\mathbf{G}(s)) = \frac{1}{d^2(s)}\{N_{11}(s)N_{22}(s) - N_{12}(s)N_{21}(s)\} = \frac{1}{C_d^2(\mathbf{G})} det(\mathbf{N}(\mathbf{G})). \tag{2.79}$$

There may be some common factors between $C_d^2(\mathbf{G})$ and $det(\mathbf{N}(\mathbf{G}))$ in (2.79), which remains in addition to roots of $p(s)$ and $z(s)$. Hence, it cannot be said that $C_d^2(\mathbf{G})$ has the same factors as $p(s)$ described in (2.78) or whether it is the least common denominator of all non-zero minors of all orders of $\mathbf{G}(s)$. However at the same time, it can definitely be said that the $p(s)$ is a factor of $C_d^2(\mathbf{G})$. Similarly, $z(s)$ is a factor of $det(\mathbf{N}(\mathbf{G}))$. Hence for detecting non-minimum phase poles and zeros of $\mathbf{G}(s)$, it is sufficient to investigate the roots of

$$C_d(\mathbf{G}) = 0 \quad \text{and} \quad det(\mathbf{N}(\mathbf{G})) = 0. \tag{2.80}$$

2.3.2 Poles and Zeros of $G_c(s)$

On basis of the conclusions of the last section, $\mathbf{G_c}(s)$, which can be represented using (2.73) as,

$$\mathbf{G_c}(s) = \frac{1}{C_d(\mathbf{G_{vof}})}\mathbf{N}(\mathbf{G_{vof}}) \times \frac{1}{C_d\left(\mathbf{G_I^{-1}}\right)}\mathbf{N}\left(\mathbf{G_I^{-1}}\right) \times \frac{1}{C_d(\mathbf{G_a})}\mathbf{N}(\mathbf{G_a}), \tag{2.81}$$

roots of the following needs to be investigated

$$\begin{aligned} C_d\left(\mathbf{G_{vof}}\right) = 0 \quad &\text{and} \quad det(\mathbf{N}\left(\mathbf{G_{vof}}\right)) = 0, \\ C_d\left(\mathbf{G_I^{-1}}\right) = 0 \quad &\text{and} \quad det(\mathbf{N}\left(\mathbf{G_I^{-1}}\right)) = 0, \\ C_d\left(\mathbf{G_a}\right) = 0 \quad &\text{and} \quad det(\mathbf{N}\left(\mathbf{G_a}\right)) = 0. \end{aligned} \tag{2.82}$$

All the per-phase transfer functions defined in (2.46)–(2.50), can be represented in the form

$$D(s) = \frac{D_Z(s)}{D_P(s)}, \tag{2.83}$$

where $D_Z(s)$ and $D_P(s)$ are the numerator and denominator polynomials respectively. Using (2.53) and (2.54), the individual elements of the corresponding transfer matrix $\mathbf{D}(s)$ in dq domain are

$$\begin{aligned} D_{11}(s) &= \frac{1}{2}\left\{\frac{D_Z(s+j\omega_T)}{D_P(s+j\omega_T)} + \frac{D_Z(s-j\omega_T)}{D_P(s-j\omega_T)}\right\}, \\ D_{12}(s) &= -\frac{j}{2}\left\{\frac{D_Z(s+j\omega_T)}{D_P(s+j\omega_T)} - \frac{D_Z(s-j\omega_T)}{D_P(s-j\omega_T)}\right\}. \end{aligned} \tag{2.84}$$

Therefore the common denominator polynomial is

$$C_d\left(\mathbf{D}\right) = D_P(s+j\omega_T)\,D_P(s-j\omega_T). \tag{2.85}$$

The roots of common denominator $C_d\left(\mathbf{D}\right)$ of the transfer matrices are obtained by shifting only the imaginary part of the roots of $D_P(s)$. Therefore the real part of the roots remain unaffected as shown in Fig. 2.9. Hence, if $D(s)$ does not have any RHP, then all the roots of $C_d\left(\mathbf{D}\right)$ lie in the left half and consequently $\mathbf{D}(s)$ cannot have a RHP. So a stable transfer function remains stable after the abc to dq transformation.

Again, determinant of $\mathbf{N}\left(\mathbf{D}\right)$ is

$$\begin{aligned} det\left(\mathbf{N}\left(\mathbf{D}\right)\right) &= D_Z(s+j\omega_T)D_P(s-j\omega_T) \\ &\quad \times D_Z(s-j\omega_T)D_P(s+j\omega_T). \end{aligned} \tag{2.86}$$

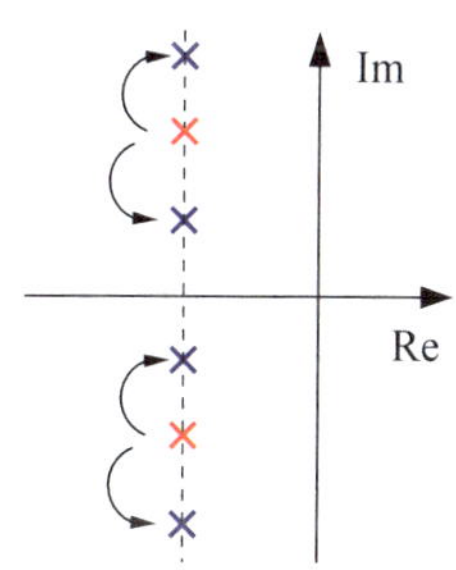

Fig. 2.9 Roots of $D(s)$ and $C_d\left(\mathbf{D}\right)$

So, if $D(s)$ do not have any RHP or RHZ, then $det\ (\mathbf{N})$ cannot have any root with positive real part either and hence $\mathbf{D}(s)$ cannot have a RHZ. Therefore, since $Z_{so}(s)$, $G_{vof}(s)$ and $Y_{mo}(s)$ do not have any RHP or RHZ, the corresponding transfer matrices in dq domain i.e. $\mathbf{Z_{so}}(s)$, $\mathbf{G_{vof}}(s)$ and $\mathbf{Y_{mo}}(s)$ do not contain any RHP/RHZ either. Therefore $C_d\ (\mathbf{Z_{so}})$, $C_d\ (\mathbf{Y_{mo}})$ and $C_d\ (\mathbf{G_{vof}})$ do not have any root in the right half of s plane.

Now the poles and zeros of the two component matrices $\mathbf{G_I^{-1}}(s)$ and $\mathbf{G_a}(s)$ will be examined.

2.3.2.1 Poles of $\mathbf{G_I^{-1}}(s)$ and $\mathbf{G_a}(s)$

Expression for the common denominator of $\mathbf{G_I^{-1}}(s)$ has been derived in (A.18) as

$$C_d\left(\mathbf{G_I^{-1}}(s)\right) = C_d\left(\mathbf{Y_{mo}}(s)\right) C_d\left(\mathbf{Z_{so}}(s)\right)\left(1 + \underbrace{\frac{3}{4}m^2 Y_{mo11}(s) Z_{so11}(s)}_{G2}\right). \tag{2.87}$$

$C_d\ (\mathbf{Y_{mo}}(s))\ C_d\ (\mathbf{Z_{so}}(s))$ do not have any root with positive real part. To reduce losses, all input filters are designed to be under damped. Consequently, $Z_{so}(s)$ in (2.46) can be represented with reasonably high accuracy for all frequencies as

$$Z_{so}(s) = \frac{s(L_s + L_f)}{s^2(L_s + L_f)C_f + s\dfrac{L_f^2}{R_d(L_s + L_f)} + 1}. \tag{2.88}$$

In terms of corner frequency (ω_c) and quality factor (Q), $Z_{so}(j\omega)$ can be represented as

$$Z_{so}(j\omega) = \left(L_s + L_f\right)\frac{j\omega}{\left(1 - \dfrac{\omega^2}{\omega_c^2}\right) + j\dfrac{\omega}{Q\omega_c}}, \tag{2.89}$$

where

$$\omega_c \approx \frac{1}{\sqrt{(n+1)L_f C_f}},\ Q \approx Rd\left(\sqrt{\frac{C_f}{L_s}}\right)(1+n)^{1.5}\ \&\ n = \frac{L_s}{L_f}. \tag{2.90}$$

Denoting the following

$$r_\omega = \frac{\omega}{\omega_c}\quad \&\quad K = L_s + L_f, \tag{2.91}$$

$Z_{so}(j\omega)$ in (2.89) can be expressed as

$$Z_{so}(j\omega) = K\omega_c \frac{jr_\omega}{(1 - r_\omega^2) + j\frac{r_\omega}{Q}} = K\omega_c \frac{jr_\omega \left\{(1 - r_\omega^2) - j\frac{r_\omega}{Q}\right\}}{(1 - r_\omega^2)^2 + \frac{r_\omega^2}{Q^2}}$$

$$= Pj\left\{(1 - r_\omega^2) - j\frac{r_\omega}{Q}\right\} = P\left\{\frac{r_\omega}{Q} + j\left(1 - r_\omega^2\right)\right\}, \qquad (2.92)$$

where P is a positive number. Therefore

$$\Re\{Z_{so}(j\omega)\} > 0 \quad \forall\, \omega. \qquad (2.93)$$

At the output side for any kind of load, virtual damping can be introduced which would be discussed in a later chapter. So the quality factor of the output filter can be controlled. Consequently, for any kind of load, it can be shown in a very similar manner as done for $Z_{so}(j\omega)$ that

$$\Re\{Y_{mo}(j\omega)\} \geq 0 \quad \forall\, \omega. \qquad (2.94)$$

Therefore, using (2.84)

$$\Re\{Z_{so11}(j\omega)\} > 0 \quad \text{and} \quad \Re\{Y_{mo11}(j\omega)\} \geq 0. \qquad (2.95)$$

Hence denoting the phase of both $Z_{so11}(j\omega)$ and $Y_{so11}(j\omega)$ as $\angle Z_{so11}(j\omega)$ and $\angle Y_{mo11}(j\omega)$ respectively, the following can be stated

$$-90^\circ < \angle Z_{so11}(j\omega) < 90^\circ \quad \& \quad -90^\circ \leq \angle Y_{mo11}(j\omega) \leq 90^\circ, \qquad (2.96)$$

and therefore,

$$-180^\circ < \angle Z_{so11}(j\omega)Y_{mo11}(j\omega) < 180^\circ. \qquad (2.97)$$

Hence, plot of $Y_{mo11}(j\omega)Z_{so11}(j\omega)$, for all possible values of filter parameters and load, never encircles the $(-1 + j0)$ point as shown in Fig. 2.10. Therefore from

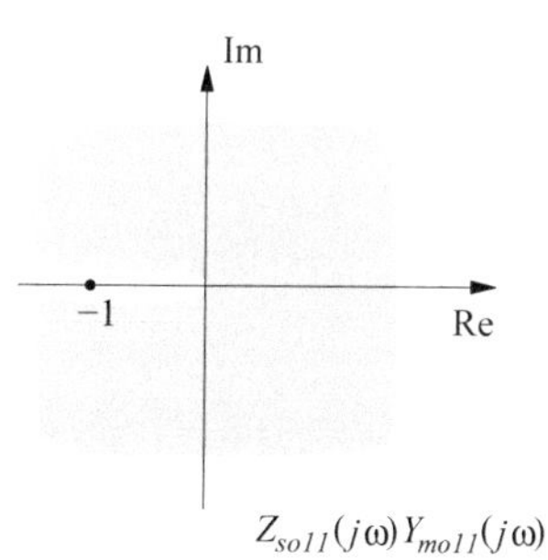

Fig. 2.10 Shaded area encompassing all possible plots of $Y_{mo11}(j\omega)Z_{so11}(j\omega)$

Nyquist criterion it can be concluded that in (2.87), $(1 + G2)$ does not have a root with positive real part. Therefore $\mathbf{G_I^{-1}}(s)$ does not contain a RHP.

$\mathbf{G_a}(s)$ cannot have a RHP either, as from Appendix A.3, its common denominator is

$$C_d\left(\mathbf{G_a}(s)\right) = C_d\left(\mathbf{Z_{so}}(s)\right). \tag{2.98}$$

Since $\mathbf{G_I^{-1}}(s)$, $\mathbf{G_{vof}}(s)$ and $\mathbf{G_a}(s)$ do not contain any RHP, it is therefore concluded that $\mathbf{G_c}(s)$ cannot have any RHP.

2.3.2.2 Zeros of $\mathbf{G_I^{-1}}(s)$ and $\mathbf{G_a}(s)$

Recalling that

$$\mathbf{G_c}(s) = \mathbf{G_{vof}}(s)\mathbf{G_I^{-1}}(s)\mathbf{G_a}(s), \tag{2.99}$$

determinant of $\mathbf{N}\left(\mathbf{G_I^{-1}}(s)\right)$ has been derived in Appendix A.3 as

$$det\left(\mathbf{N}\left(\mathbf{G_I^{-1}}(s)\right)\right) = C_d\left(\mathbf{G_I^{-1}}(s)\right) C_d\left(\mathbf{Y_{mo}}(s)\right) C_d\left(\mathbf{Z_{so}}(s)\right). \tag{2.100}$$

It has been discussed in the last section that $C_d\left(\mathbf{G_I^{-1}}(s)\right)$ does not contain any right half zeros and neither does $C_d\left(\mathbf{Y_{mo}}(s)\right)$ nor $C_d\left(\mathbf{Z_{so}}(s)\right)$. Hence, $\mathbf{G_I^{-1}}(s)$ can not have any RHZ.

So, now focusing on the zeros of $\mathbf{G_a}(s)$, from (A.21)

$$det\left(\mathbf{N}\left(\mathbf{G_a}(s)\right)\right) = C_d^2\left(\mathbf{Z_{so}}(s)\right) a_V^2 \left[1 - \frac{M_d I_{od} + M_q I_{oq}}{a_V} Z_{so11}(s)\right], \tag{2.101}$$

where,

$$\begin{bmatrix} I_{od} \\ I_{oq} \end{bmatrix} = \sqrt{\frac{3}{2}}\,\hat{I}_o \begin{bmatrix} \cos(\varphi_o + \varphi_{oL}) \\ \sin(\varphi_o + \varphi_{oL}) \end{bmatrix}. \tag{2.102}$$

Using (2.27), (2.35), (2.44) and (2.102)

$$M_d I_{od} + M_q I_{oq} = \sqrt{\tfrac{3}{2}}\,\hat{I}_{in}. \tag{2.103}$$

Substituting (2.103) in (2.101),

$$det\left(\mathbf{N}\left(\mathbf{G_a}(s)\right)\right) = C_d^2\left(\mathbf{Z_{so}}(s)\right) a_V^2 \left[1 - \frac{\sqrt{\frac{3}{2}}\,\hat{I}_{in}}{a_V} Z_{so11}(s)\right]. \tag{2.104}$$

Thereafter using (2.25) along with (2.71) in (2.104),

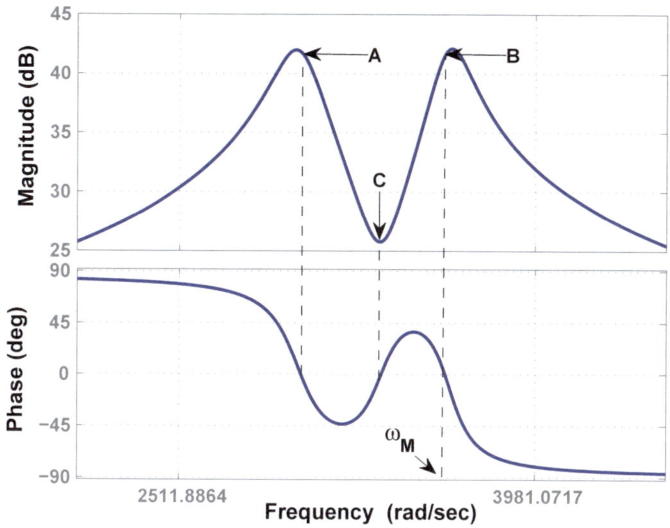

Fig. 2.11 Frequency response of $Z_{so11}(s)$

$$det(\mathbf{N}(\mathbf{G_a}(s))) = C_d^2(\mathbf{Z_{so}}(s))\, a_V^2 \left[1 - \frac{P_{in}}{\frac{3}{2}\hat{V}_{cf}^2 \cos^2 \varphi_i} Z_{so11}(s)\right]$$
$$= C_d^2(\mathbf{Z_{so}}(s))\, a_V^2 \left[1 - \frac{1}{R_n} Z_{so11}(s)\right]. \tag{2.105}$$

Frequency response of $Z_{so11}(s)$ is shown in Fig. 2.11. The magnitude plot shows two adjacent peaks near the input filter corner frequency. Also, multiple zero crossing of phase plot is observed. The frequency of the zero phase corresponding to the maximum magnitude is chosen. For a given set of parameters, let point B be that maximum, as shown in Fig. 2.11, corresponding to frequency ω_M. Owing to the presence of negative sign preceding $(1/R_n)\, Z_{so11}$ in (2.105), sufficient condition for avoiding RHZ in $\mathbf{G_a}$, from Nyquist criterion, is

$$\frac{P_{in}}{\frac{3}{2}\hat{V}_{cf}^2 \cos^2 \varphi_i} |Z_{so11}(j\omega_M)| < 1, \quad \text{or} \quad P_{in} < \frac{\frac{3}{2}\hat{V}_{cf}^2 \cos^2 \varphi_i}{|Z_{so11}(j\omega_M)|} = P_{cr}. \tag{2.106}$$

So, emergence of RHZ depends on the magnitude of voltage at the input terminals, input power factor, power transfer, input filter parameters and source inductance. Moreover, $||Z_{so11}(j\omega)||_\infty$ increases with L_s/L_f ratio and so does $|Z_{so11}(j\omega_M)|$. Hence a higher L_s brings down the P_{in} limit (P_{cr}) for RHZ.

It is noted that the present modelling is achieved without considering the parasitic resistances of the filter elements which increase the overall damping in an actual plant. Hence, strictly following (2.106) would lead to conservative design. In this context, the parasitic resistances for the inductors depends on the construction, wire gauge etc. which makes it very difficult to address all these aspects in a generic model.

Nevertheless, this analysis gives the system designer an insight into the plant model and a rule to approach the design in an integrated manner. There are two choices—either working with a minimum phase system by appropriate input filter design or designing a suitable controller for a non-minimum phase plant. The former choice simplifies the controller design. However if the load requirement makes the later situation inevitable, then, a proper controller design ensuring stable closed loop operation becomes imperative.

2.4 Concluding Remarks

A model of a grid connected 3ϕ MC has been derived incorporating input, output filters and source inductance. It has been clearly shown that based on a given operating point, RHZs may appear in the control to output transfer function which depends upon the input filter parameters. The power stage of MC along with the presence of input filter is responsible for the emergence of these RHZs in the linearized model. This is clarified with the following example of a generic DC-DC converter having a power stage equivalent to MC.

Example: Figure 2.12 shows the a DC-DC converter where the control objective is to regulate i_o. The input and output voltages and currents are related as

$$v_{on} = dv_{cf} \quad \& \quad i_{in} = di_o = \frac{d^2 v_{cf}}{R} \tag{2.107}$$

where d, the duty cycle is the control input. The dynamic equation of the state v_{cf} is

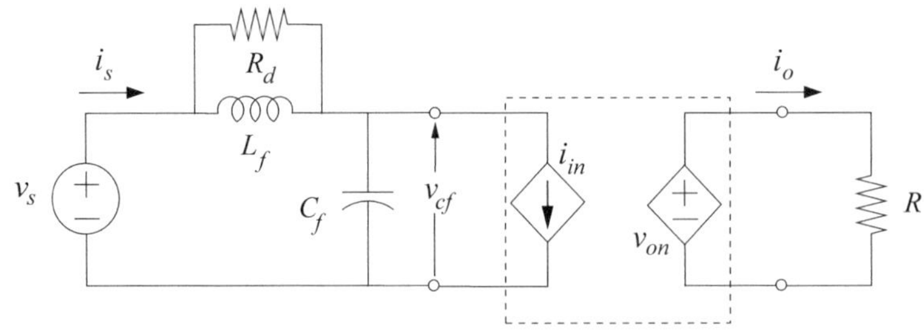

Fig. 2.12 Generic DC-DC converter having same power stage as MC

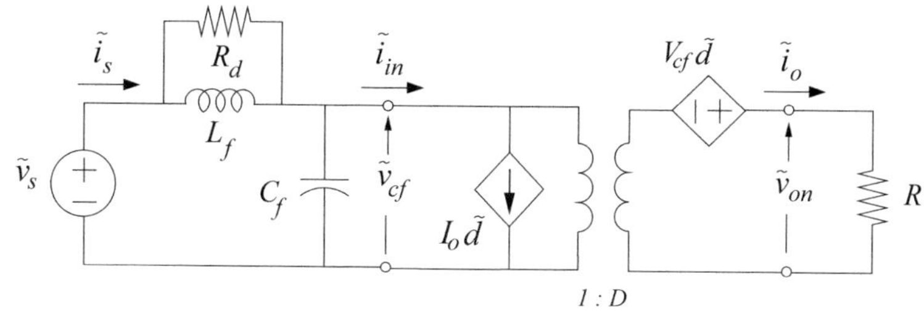

Fig. 2.13 Equivalent circuit of the linearized system

$$\frac{dv_{cf}}{dt} = \frac{1}{C_f}\left(i_s - \frac{d^2 v_{cf}}{R}\right). \tag{2.108}$$

Due to product of the state and control input in (2.108), linearization becomes necessary. In the linearized model shown in Fig. 2.13, the control transfer function is obtained as

$$\frac{\tilde{i}_o(s)}{\tilde{d}(s)} = \frac{V_{cf}}{R}\frac{s^2 L_f C_f + s L_f\left(\frac{1}{R_d} - \frac{D^2}{R}\right) + 1}{s^2 L_f C_f + s L_f\left(\frac{1}{R_d} + \frac{D^2}{R}\right) + 1}. \tag{2.109}$$

Hence RHZs emerge if,

$$\frac{R}{D^2} < R_d, \tag{2.110}$$

i.e. when the load resistance referred to source side is less than the damping resistor. Again since,

$$\frac{R}{D^2} = \frac{V_{on}}{I_o D^2} = \frac{V_{cf}}{I_{in}} = \frac{V_{cf}^2}{P_{in}}, \tag{2.111}$$

therefore, the condition for RHZs described in (2.110), can also be represented as

$$\frac{P_{in}}{V_{cf}^2} > \frac{1}{R_d}. \tag{2.112}$$

Meeting the control requirement i.e. regulation of i_o leads to the origin of the RHZ. From the small signal model, the input side and the output side variables are related as

$$\tilde{v}_{on}(s) = D\tilde{v}_{cf}(s) + V_{cf}\tilde{d}(s),$$
$$\tilde{i}_{in}(s) = D\tilde{i}_o(s) + I_o\tilde{d}(s). \tag{2.113}$$

Tightly regulated i_o means that irrespective of the variation in v_s, the power transferred to the load P_{in} remains constant. In other words, perfect control demands the following

$$\tilde{v}_{on}(s) = 0, \ \tilde{i}_o(s) = 0. \tag{2.114}$$

Using (2.113) and (2.114) results in the following

$$\frac{\tilde{v}_{cf}(s)}{\tilde{i}_{in}(s)} = -\frac{V_{cf}}{DI_o} = -\frac{V_{cf}}{I_{in}} = -\frac{R}{D^2} \tag{2.115}$$

Therefore, under the assumption of perfect regulation the load appears as a negative resistance in the linear model. Any increment in $\tilde{v}_{cf}$ is met with a simultaneous decrement in $\tilde{i}_{in}$, thus exhibiting the typical RHZ characteristics.

The presented analysis is applicable to any other modulation or modelling technique. Thus a design constraint for overall system design has been established which will be used in subsequent chapters where filter and controller design has been detailed.

Chapter 3
Filter Design

This chapter details design of input and output filters for a 3-Ph MC. The design method targets limitation of conducted electromagnetic interference (EMI) only. The discussed methodology is not devised to meet any specific EMI standard. This method evolves by setting individual targets for specific performance parameters and subsequently combining them to identify an allowable solution set which ensures compliance to all the specifications. To begin with, a set of performance criteria for the input filter is stated, based on switching frequency ripple attenuation, voltage regulation and input reactive current. Instantaneous input-output power balance in MC mandates the following

$$\hat{V}_{cf}\hat{I}_{in}\cos\varphi_i = \hat{V}_o\hat{I}_o\cos\varphi_{oL}. \tag{3.1}$$

Since in most applications the control variables are v_o and $\cos\varphi_i$, there is no scope for controlling the input current amplitude. Therefore damping of any oscillations in the input filter has to be provided by including an external damping resistor. Approximation of the power loss due to this inclusion is presented. Note that the internal resistances of the different elements are not considered here and therefore the damping resistor represents the only power dissipating element.

Apart from this, certain constraints also arise due to operational characteristics which are specific to MC. It was discussed in Chap. 1 that commutation in MC has to be carefully orchestrated such that the switching rules are not violated. The working principles of the two commutation methods were also discussed. Output voltage regulation requirements limit the size of the output filter inductor (filter component in series) leading to a high ripple in the output current through the inductor. This makes commutation based on zero crossing (ZC) of current difficult, hence voltage based commutation has been adopted for the applications considered in this book.

A. Dasgupta and P. Sensarma, *Design and Control of Matrix Converters*,
Energy Systems in Electrical Engineering, DOI 10.1007/978-981-10-3831-0_3

In the subsequent section, input filter design in presented. Input capacitor sizing—for reliable commutation—is based on the voltage ripple and consequent problems in voltage based commutation. The criterion for a minimum phase plant is revisited and the consequent effects on input filter performance and controller design are discussed. Thereafter, in Sect. 3.2, output filter design is presented where sizing of the corresponding filter elements has been discussed. Thus a system design approach emerges which helps to meet both steady state and dynamic performance requirements. The entire filter design procedure is summarized in Sect. 3.3 to aid a structured understanding. Finally, relevant experimental results on a 6 kVA experimental prototype are provided for validation.

3.1 Input Filter Design

Filter parameters are chosen to satisfy certain steady state specifications for the nominal system parameters shown in Table 3.1.

Figure 3.1 shows the per phase equivalent circuit for the input filter of MC. Internal (Thevenin) impedance of the source is modelled as an inductance L_s. Damping resistor, R_d, is placed in parallel with the filter inductor, which best eases the conflict between filter efficiency and damping [46]. The detailed design approach is presented neglecting L_s since it is usually small in distribution systems. Once the filter parameters are selected, L_s is introduced to study how the overall filter performance gets affected by its addition.

The selection of filter parameters are sequentially described in the following sections.

Table 3.1 Nominal/rated parameters

Base frequency (f_b)	Source voltage (per phase) (V_s)	Switching frequency (f_s)	Output current (per phase) (I_o)
50 Hz	240 V	10 kHz	10 A

Fig. 3.1 Input filter

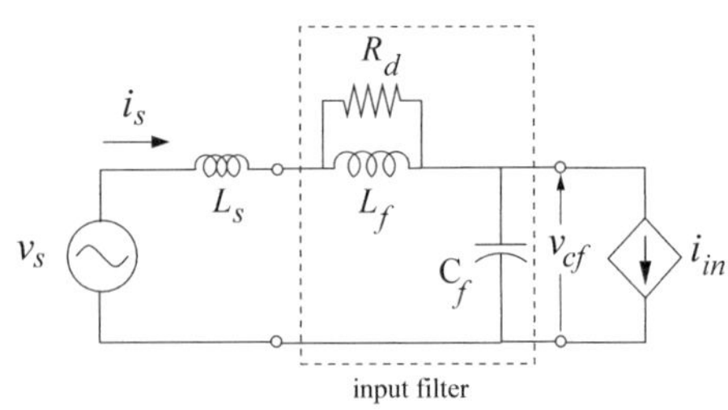

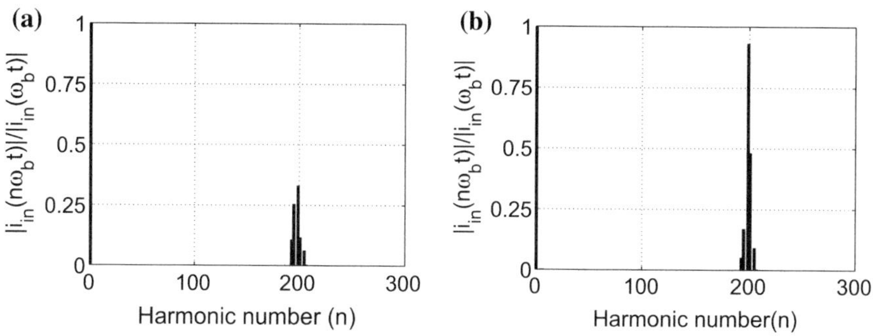

Fig. 3.2 $|i_{in}(n\omega_b t)|/|i_{in}(\omega_b t)|$ for **a** m=1, **b** $m = 0.5$

3.1.1 Attenuation to Switching Frequency Ripple and Gain to Lower Order Harmonics

For $L_s = 0$, the forward gain of input filter defined in (2.49), is obtained as

$$G_{fv}(s) = \left.\frac{i_s(s)}{i_{in}(s)}\right|_{v_s} = \left.\frac{v_{cf}(s)}{v_s(s)}\right|_{i_{in}} = \frac{s\frac{L_f}{R_d} + 1}{s^2 L_f C_f + s\frac{L_f}{R_d} + 1}. \tag{3.2}$$

Denoting the corner frequency as ω_c ($=2\pi f_c$) and quality factor as Q, the amplitude of its frequency response is expressed in a normalized form as

$$|G_{fv}(j\omega)| = \sqrt{\left[1 + \frac{r_\omega^2}{Q^2}\right] / \left[(1 - r_\omega^2)^2 + \frac{r_\omega^2}{Q^2}\right]}, \tag{3.3}$$

and plotted in Fig. 3.3. The definitions

$$r_\omega = \frac{\omega}{\omega_c}, \quad Q = R_d\sqrt{\frac{C_f}{L_f}}, \quad \omega_c = \frac{1}{\sqrt{L_f C_f}} = 2\pi f_c, \tag{3.4}$$

enable generalized analysis.

Magnitude ratio of the switching ripple and fundamental components in i_{in} varies with modulation index, m [47], which is also observed in the simulation results for the first (dominant) sideband cluster shown in Fig. 3.2.

To ensure acceptable switching ripple content in the source current, i_s, the first design criterion is therefore set as,

Spec.1 $\quad |G_{fv}(j2\pi f_s)| \leq A_{sw}$ dB.

Low order harmonics in the input current could also occur because of distorted distribution grid voltages at the site where the MC is installed. Denoting the corresponding low-order grid voltage harmonic order as h_v, a limit on the amplification of these in the filter capacitor voltage, v_{cf}, is introduced by the following restriction.

Spec.2 $|G_{fv}(j2\pi f_b m_{fhv})| \le A_{vh}$ dB.

where,

$$m_{fhv} = \max.(h_v) \tag{3.5}$$

is the maximum significant harmonic order in the spectrum of the source voltage, v_s. Once the switching frequency, f_s, is decided and the deployment site conditions, signified by the harmonic orders of the grid voltage, are known, the following frequency ratio, defined as

$$m_{fsh} = \frac{f_s}{m_{fhb} f_b}, \tag{3.6}$$

has a unique value. The horizontal lines in Fig. 3.3 depict the limits defined in **Spec.1** and **Spec.2**. These lines intersect each individual curve, drawn for a distinct value of Q, at two separate points. As an example, these points are marked as A and B, for the curve corresponding to $Q = 5$. Let the normalized frequencies, at these points of intersection, be $r_{\omega v}$ and $r_{\omega s}$. Defining the ratio, for $Q = Q_k$, as

$$m_{hs}^{(k)} = \frac{r_{\omega s}}{r_{\omega v}}, \; Q = Q_k \tag{3.7}$$

a valid choice of Q_k must satisfy the relation

$$m_{hs}^{(k)} \le m_{fsh}. \tag{3.8}$$

The range of $\{Q : Q^{min} \le Q \le Q^{max}\}$ for which (3.8) is satisfied defines the solution space for Q. For any value of $Q = Q^{(1)}$ within the open range $Q^{min} < Q^{(1)} < Q^{max}$, the acceptable range of $\{f_c : f_{c,\,\mathrm{min}}^{(1)} \le f_c \le f_{c,\,\mathrm{max}}^{(1)}\}$ is determined. For the experimental model and grid parameters at the site of experiment, these selections, along with the numerical limits of **Spec.1** and **Spec.2**, are listed in Table 3.2.

Note that if (3.8) is not satisfied, then it is not possible to find a feasible solution with the given site constraints, limiting values and choice of switching frequency. A simple way to overcome this impasse would be to increase the value of switching frequency.

Table 3.2 Q and f_c from forward gain

A_{sw}	$\max.(h_v)$	A_{vh}	$Q^{(1)}$	$f_{c,\,\mathrm{max}}^{(1)}$	$f_{c,\,\mathrm{min}}^{(1)}$
−26 dB	7	2 dB	3	1.35 kHz	772 Hz

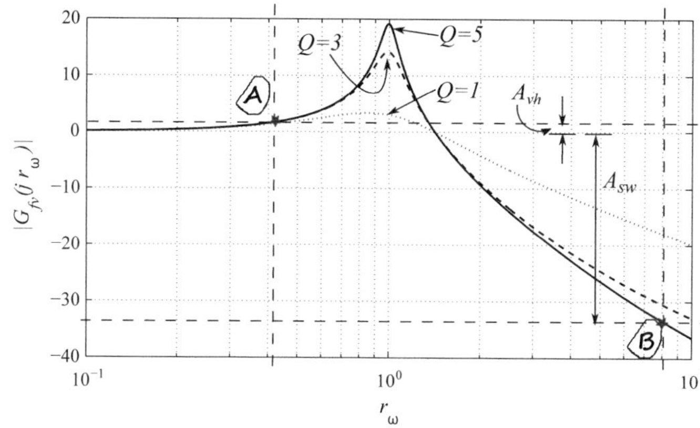

Fig. 3.3 $|G_{fv}(jr_\omega)|$(dB) with $Q = 1$, 3 and 5

3.1.2 Voltage Regulation and Reactive Current Loading

Value of the filter inductance L_f decides the fundamental voltage drop across the filter, which determines the full load regulation. A significant portion of the fundamental frequency component of input current i_s has to flow through L_f to restrict the losses in R_d. The initial choices of $Q^{(1)}$ and $[f_{c,\min}^{(1)}, f_{c,\max}^{(1)}]$ automatically ensures that. This can be appreciated from the following relations

$$Q = R_d\sqrt{\frac{C_f}{L_f}} = \frac{R_d}{L_f\omega_c} \Longrightarrow R_d = QL_f\omega_c = Q\left(\frac{\omega_c}{\omega_b}\right)\omega_b L_f. \tag{3.9}$$

Using the numerical values of $Q^{(1)}$ and $[f_{c,\min}^{(1)}, f_{c,\max}^{(1)}]$ from Table 3.2, it can be concluded that

$$R_d \geq 46\omega_b L_f. \tag{3.10}$$

Therefore I_{L_f} can be expressed as

$$I_{L_f} \approx I_s = \sqrt{I_{cf}^2 + I_{in}^2} \tag{3.11}$$

This restriction on voltage regulation leads to another important design consideration.

Spec.3 $V_{Lf} \approx \omega_b L_f \sqrt{I_{cf}^2 + I_{in,\ \text{rated}}^2} \leq k_{\mathcal{R}} V_s.$

Utility grids usually specify a minimum leading power factor, typically occurring at part-load conditions. This leads to a restriction on reactive current loading, which is formulated as the next design criterion.

Table 3.3 $L_{f,\,\max}$ and $C_{f,\,\max}$ from regulation and reactive loading

$k_{\mathcal{PF}}$	$k_{\mathcal{R}}$	$L^{(1)}_{f,\,\max}$	$C^{(1)}_{f,\,\max}$
0.2	0.03	2.6 mH	22 μF

Spec.4 $\quad I_{cf} \approx \omega_b C_f V_s \leq k_{\mathcal{PF}} I_{in,\,\text{rated}}.$

For deriving the boundary values at rated power level, analysis is done considering unity IDF operation with unity modulation index (m) [25] and resistive load unless stated otherwise. The specific inequalities in **Spec.3** and **Spec.4** obviously translate to maximum limits for L_f ($L^{(1)}_{f,\,\max}$) and C_f ($C^{(1)}_{f,\,\max}$). The rated input current is evaluated from [25]

$$I_{in,\,\max} = 0.866 I_o. \tag{3.12}$$

For the rated values in Table 3.1, numerical limits considered for **Spec.3** and **Spec.4** along with the maximum allowable values $L^{(1)}_{f,\,\max}$ and $C^{(1)}_{f,\,\max}$ are listed in Table 3.3. Note that these values are not sacrosanct and can be chosen differently for systems where the nominal power level is different from what has been considered here.

3.1.3 Selecting Damping Resistor R_d

Introducing the external damping resistor R_d invariably adds to the losses in input filter. So R_d has to be chosen judiciously such that the efficiency does not degrade beyond an acceptable limit. To address this, a limit has to be placed on the maximum loss occurring in R_d. Denoting the current through R_d as i_{R_d}, the corresponding design criterion is stated as

Spec.5 $\quad 3R_d \sum_{n=1}^{\infty} [i_{R_d}(jn\omega_b)]^2 < 1\%$ of rated load.

Now, i_{R_d} is dependent on both the input voltage, v_s, and the input current, i_{in}, of the MC. This dependence is expressed as

$$i_{R_d}(s) = G_{R1}(s) v_s(s) + G_{R2}(s) i_{in}(s), \tag{3.13}$$

where,

$$\left| G_{R1}(s) \triangleq \frac{i_{R_d}(s)}{v_s(s)} \right|_{s=j\omega} = \frac{1}{R_d} \cdot \frac{r_\omega^2}{\sqrt{\left[(1-r_\omega^2)^2 + \frac{r_\omega^2}{Q^2}\right]}},$$

$$\left| G_{R2}(s) \triangleq \frac{i_{R_d}(s)}{i_{in}(s)} \right|_{s=j\omega} = \frac{1}{Q} \cdot \frac{r_\omega}{\sqrt{\left[(1-r_\omega^2)^2 + \frac{r_\omega^2}{Q^2}\right]}}. \tag{3.14}$$

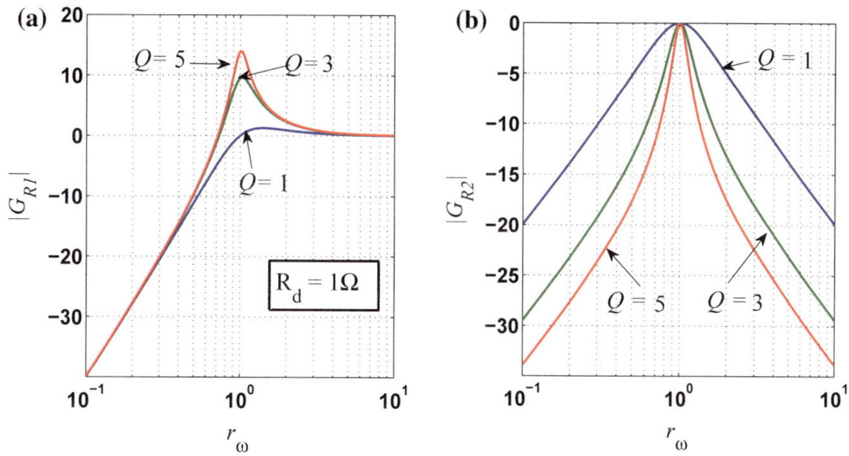

Fig. 3.4 **a** $|G_{R1}(j\omega)|$(dB) for $R_d = 1\Omega$, **b** $|G_{R2}(j\omega)|$(dB)

Fig. 3.4a, b plots the frequency response of G_{R1} and G_{R2}, respectively, for $R_d = 1\Omega$.

A high value of Q is actually desirable in $|G_{R2}|$ since this implies lesser ripple component in i_{R_d}, hence lower loss. But, a similar consideration (high Q) in $|G_{R1}|$ indicates higher loss due to increase in low order harmonic content in i_{R_d}. These plots exhibit response similar to that experienced with the normalized form of $|G_{fv}|$ in Fig. 3.3. From this similarity, $Q^{(1)}$ chosen in Sect. 3.1.1 expectedly leads to an even balance between both these conflicting aspects. Thus, the value of $Q^{(1)}$ is finalized for the input filter.

Power loss occurring in R_d is given by

$$P_R = 3R_d \left\{ \sum_n \left(|G_{R1}(jn\omega_b)|^2 V_{s,n}^2 + |G_{R2}(jn\omega_b)|^2 I_{in,n}^2 \right) \right\}, \qquad (3.15)$$

where, $V_{s,n}$ and $I_{in,n}$ are the r.m.s values of the nth harmonic order in v_s and i_{in}, respectively. Calculation of P_R using (3.15) requires prior selection of f_c for evaluating the numerical values of the gains $|G_{R1}(jn\omega_b)|$ and $|G_{R2}(jn\omega_b)|$. For the experimental model, a selection of f_c is made from within the allowable range listed in Table 3.2 as

$$f_c^{(1)} = 1 \text{ kHz}. \qquad (3.16)$$

3.1.3.1 R_d from Loss Limit

An estimate of P_R is derived as follows, assuming sinusoidal i_o. A simulation of the MC system is carried out and the harmonic content observed in i_{in} are listed in

Table 3.4 Harmonics in i_{in} from simulation at $m = 1$

193rd	195th	199th	201st	205th	207th
10.66 %	25.31 %	32.58 %	11.88 %	5.84 %	2.58 %

Table 3.5 F1 and F2

F_1	F_2
3.44	$0.001167I_o^2$

Table 3.4. Also, numerical values of the gains defined in (3.14) are evaluated using $f_c^{(1)}$ and $Q^{(1)}$. These gain values and the harmonic content information, listed in Table 3.4, are inserted in (3.15), along with the nominal parameters in Table 3.1, to obtain P_R in the following form.

$$P_R = \frac{F_1}{R_d} + F_2 R_d \tag{3.17}$$

where F_1 and F_2 are functions of $|G_{R1}(j\omega_b)|$, V_s, $|G_{R2}(jn\omega_b)|$ and $I_{in,n}$ and are listed in Table 3.5.

Considering the numerical limit of P_R, as specified in **Spec.5**, in (3.17), a quadratic equation is R_d is obtained. Assuming both its roots are positive, these define an allowable range for R_d as follows.

$$R_{d,\,\min}^{(1)} \leq R_d \leq R_{d,\,\max}^{(1)}. \tag{3.18}$$

This choice of R_d definitely satisfies **Spec.1**, **Spec.2** and **Spec.5**. What is not apparent, however, is whether any value of R_d in this range and the corresponding L_f and C_f designed using (3.4) also satisfy the voltage regulation and maximum leading power factor criteria of **Spec.3** and **Spec.4**.

3.1.3.2 R_d from Voltage Regulation and Leading Reactive Power Factor Limits

Since the value of f_c is already decided by (3.16), the upper bounds $C_{f,\,\max}^{(1)}$ and $L_{f,\,\max}^{(1)}$, mentioned in Table 3.3, directly lead to their corresponding lower bounds $L_{f,\,\min}^{(1)}$ and $C_{f,\,\min}^{(1)}$. These are listed in Table 3.6. From (3.4),

Table 3.6 $L_{f,\,\min}$ and $C_{f,\,\min}$ from regulation, reactive loading and fixed f_c

$L_{f,\,\min}^{(1)}$	$C_{f,\,\min}^{(1)}$	$R_{d,\,\min}^{(1)}$	$R_{d,\,\max}^{(1)}$	$R_{d,\,\min}^{(2)}$	$R_{d,\,\max}^{(2)}$
1.15 mH	9.7 μF	0.0175 Ω	534.16 Ω	21.7 Ω	45 Ω

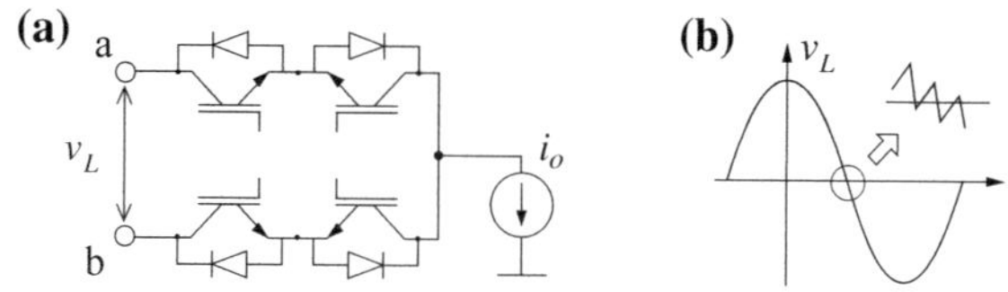

Fig. 3.5 Commutation between input phases (a, b)

$$R_d = \left(2\pi f_c^{(1)} Q^{(1)}\right) L_f \tag{3.19}$$

which leads to another set of bounds on R_d, on substituting the range limits $L_{f,\,\max}^{(1)}$ and $L_{f,\,\min}^{(1)}$ in (3.19). These bounds are expressed as

$$R_{d,\,\min}^{(2)} \le R_d \le R_{d,\,\max}^{(2)}. \tag{3.20}$$

Obviously, the allowable values of R_d lie in the intersection of the two ranges obtained from (3.18) and (3.20). If this intersection is a nullity, numerical limits for all or some of **Spec.3**, **Spec.4** and **Spec.5** must be relaxed. For the experimental model, these ranges of R_d are listed in Table 3.6, which, incidentally, did intersect and so (3.20) defined the final set for selecting R_d.

3.1.4 Lower Limit of C_f

Figure 3.5a shows the input phases-a and b, which are to be alternatively switched to the same output phase carrying output current i_o. Voltage commutation between these two phases is based on the polarity of the line-line voltage ($v_L = v_{ab}$) on the input side. Exact measurement of its polarity becomes difficult at the ZC of v_L, since the switching ripple component, v_{rL}, shown in Fig. 3.5b, causes multiple zero crossings.

In this analysis it is assumed that the spectra of both the input current and filter capacitor voltages exclusively contain fundamental and switching frequency components only. For phases-a and b, the line-line voltage is denoted as

$$\begin{aligned} v_L = v_{ab} &= (v_{fa} + v_{ra}) - (v_{fb} + v_{rb}) \\ &= \underbrace{(v_{fa} - v_{fb})}_{v_{fL}} + \underbrace{(v_{ra} - v_{rb})}_{v_{rL}}, \end{aligned} \tag{3.21}$$

where, the pair (v_{fa}, v_{ra}) denote the fundamental and ripple components of the input filter capacitor voltage of phase-a. Similar nomenclature has been used for phase-b. Figure 3.6 shows the input current and filter capacitor voltage. It is assumed that the ripple component (i_{rx}, $x \in \{a, b, c\}$) in input current (i_x) flows only through the filter capacitor while the fundamental component (i_{fx}) circulates through the source.

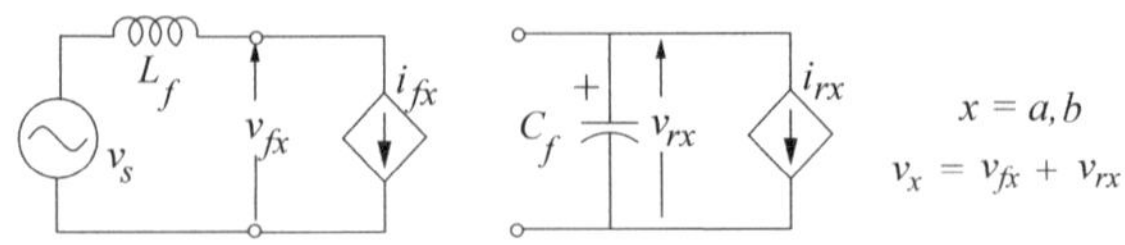

Fig. 3.6 Input current and filter capacitor voltage

Hence, v_{rL} can be expressed as,

$$v_{rL} = v_{ra} - v_{rb} = -\frac{1}{C_f}\int (i_{ra} - i_{rb})\,dt. \tag{3.22}$$

Since the MC synthesizes i_{rx} from the output current envelope, it is a direct function of the load. Hence for a fixed value of C_f, v_{rL} increases with load. Therefore an arbitrarily small C_f, chosen to reduce reactive loading at low loads, leads to a corresponding high v_{rL} at high loads. With the sensor delays involved, measurement error in v_{rL} is significantly more than that of v_{fL}. The resulting error in polarity detection of v_{rL} increases the chances of commutation failure, around ZC of v_{fL}. So, estimation of maximum v_{rL}, and investigating its effect on commutation, are essential for determination of the lower limit of C_f. This estimation requires revisiting the modulation algorithm [25], in the following manner.

3.1.4.1 Maximum v_{rL} Between Phases-a and b

The decoupled Current source converter—Voltage source converter (CSC-VSC) construct of MC described in Chap. 2 is reproduced in Fig. 3.7a. The stationary vectors for the input current hexagon are reproduced in Fig. 3.7b. Here, stationary vector $\mathbf{I}_1$ (a,c) corresponds to the state during which, input phase-a is connected to the positive rail and phase-c to the negative rail of the imaginary DC link. In sectors II and V, switching between the two bordering stationary vectors for synthesizing the current vector $\bar{\mathbf{I}}_{\mathbf{in}}$ requires commutation between input phases-a and b. The voltage space vector $\bar{\mathbf{V}}_{\mathbf{OL}}$, can lie in any one of the sectors of output voltage hexagon shown in Fig. 3.7c. Here, $\mathbf{V}_1$ (+, −, −) indicates that the output phase-A is connected to the positive rail and both B and C are connected to the negative rail of the imaginary DC bus. Figure 3.7d, e shows the stationary vectors adjacent to $\bar{\mathbf{I}}_{\mathbf{in}}$ and $\bar{\mathbf{V}}_{\mathbf{OL}}$ described in Chap. 2. θ_{SI} and θ_{SV} are the angles between the rotating vector and the stationary vector trailing it.

Let an instant be considered when $\bar{\mathbf{I}}_{\mathbf{in}}$ lies in sector II of the input current hexagon and $\bar{\mathbf{V}}_{\mathbf{OL}}$ in sector III of the output voltage hexagon. Then the switching vectors of interest are $\mathbf{V}_2$, $\mathbf{V}_3$ and $\mathbf{I}_1$, $\mathbf{I}_2$. Therefore the four stationary vector combinations of interest are $\mathbf{V}_2\mathbf{I}_1$, $\mathbf{V}_2\mathbf{I}_2$, $\mathbf{V}_3\mathbf{I}_1$ and $\mathbf{V}_3\mathbf{I}_2$. The corresponding input-output phase connections, both in the decoupled construct and in the 3 Ph MC, are provided in Table 3.7. The fundamental and switching frequency components of the input currents in phases-a and b are shown in Fig. 3.8. Since the switching frequency is much higher than the fundamental, over one switching period, T_s, the fundamental components

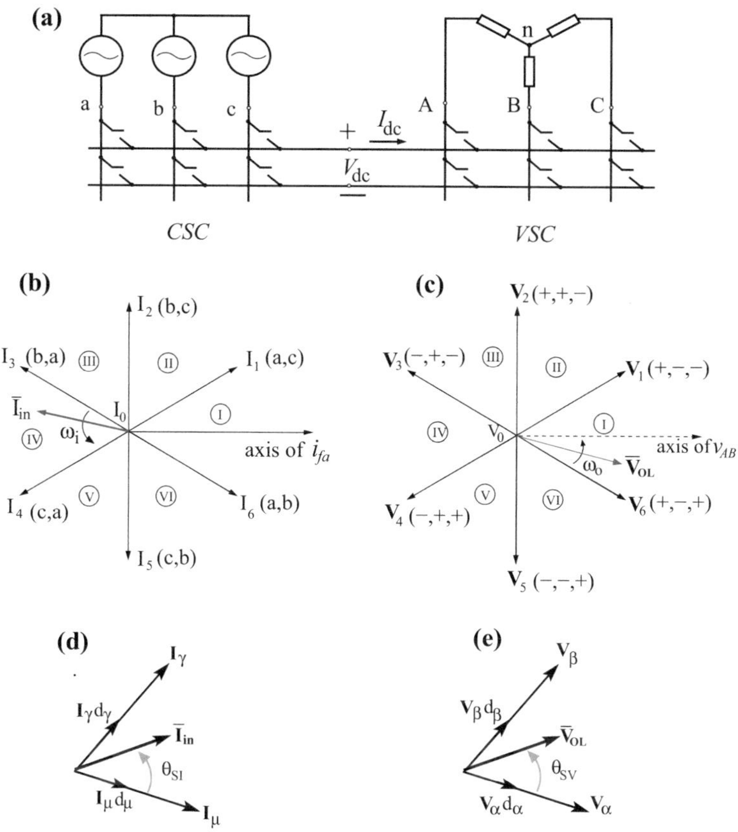

Fig. 3.7 **a** Decoupled CSC-VSC construct, **b** Input current hexagon, **c** Output voltage hexagon, **e** Synthesizing $\bar{\mathbf{I}}_{\mathbf{in}}$, **d** Synthesizing $\bar{\mathbf{V}}_{\mathbf{OL}}$

Table 3.7 Input-output connections in MC for $\bar{\mathbf{I}}_{\mathbf{in}}$ in sector II and $\bar{\mathbf{V}}_{\mathbf{OL}}$ in sector III

	ON period	Decoupled construct					3 Ph MC			i_a	i_b	i_c
		A	B	C	+	–	A	B	C			
$\mathbf{V}_2\mathbf{I}_1$	$d_\alpha d_\mu T_S \rightarrow T_{\alpha\mu}$	+	+	–	a	c	a	a	c	$-i_C$	0	i_C
$\mathbf{V}_2\mathbf{I}_2$	$d_\alpha d_\gamma T_S \rightarrow T_{\alpha\gamma}$	+	+	–	b	c	b	b	c	0	$-i_C$	i_C
$\mathbf{V}_3\mathbf{I}_1$	$d_\beta d_\mu T_S \rightarrow T_{\beta\mu}$	–	+	–	a	c	c	a	c	i_B	0	$-i_B$
$\mathbf{V}_3\mathbf{I}_2$	$d_\beta d_\gamma T_S \rightarrow T_{\beta\gamma}$	–	+	–	b	c	c	b	c	0	i_B	$-i_B$

of both input and output currents remain practically constant. For the input currents, i_{fa} and i_{fb}, these values, over a single switching period, are denoted as I_{fa} and I_{fb}, respectively. Likewise, output currents over the same interval are represented as I_C and I_B. From Table 3.7 and Fig. 3.8, using (2.30) and (2.31), I_{fa} and I_{fb} can be expressed as

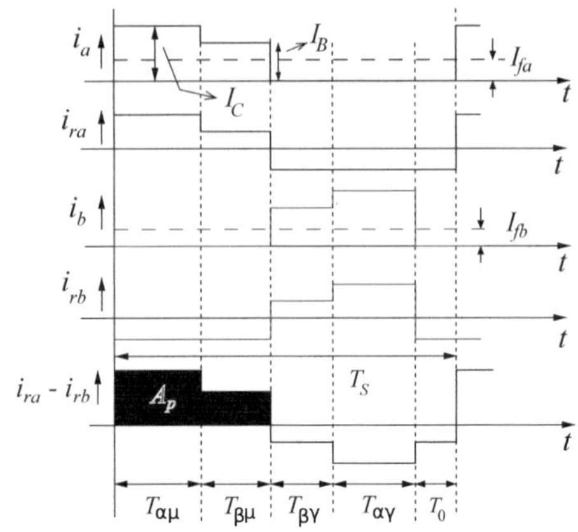

Fig. 3.8 Fundamental (I_{fa}, I_{fb}) and switching frequency (i_{ra}, i_{rb}) components of input currents (i_a, i_b)

$$\begin{aligned} I_{fa} &= \left(I_C d_\alpha d_\mu + I_B d_\beta d_\mu\right) = mI_P \sin(60^\circ - \theta_{SI}), \\ I_{fb} &= \left(I_C d_\beta d_\gamma + I_B d_\alpha d_\gamma\right) = mI_P \sin\theta_{SI}, \end{aligned} \tag{3.23}$$

where

$$I_P = I_C \sin(60^\circ - \theta_{SV}) + I_B \sin\theta_{SV}. \tag{3.24}$$

From (3.22), peak-peak magnitude of v_{rL}, $\hat{v}_{rL}$, attains its maximum value when the area A_p, shown shaded in Fig. 3.8, is maximum. A_p is evaluated as

$$A_p = \left\{(I_C - I_{fa} + I_{fb})d_\alpha d_\mu + (I_B - I_{fa} + I_{fb})d_\beta d_\mu\right\} T_s. \tag{3.25}$$

Also, i_{fa} can be expressed as

$$i_{fa} = \hat{I}_{in}\cos(\omega_i t - \theta). \tag{3.26}$$

Since $\bar{\mathbf{I}}_{\mathbf{in}}$ is in sector II of input current hexagon, using Fig. 3.7b and (3.26),

$$\theta_{SI} = \omega_i t - \theta - 30^\circ. \tag{3.27}$$

Using (3.27) to substitute $(\omega_i t - \theta)$ in (3.26) results in

$$i_{fa} = \hat{I}_{in}\sin(60^\circ - \theta_{SI}). \tag{3.28}$$

Comparing (3.28) with I_{fa} described in (3.23) gives

$$mI_P = \hat{I}_{in}. \tag{3.29}$$

Subsequently using (3.23) and (3.29), A_P in (3.25) gets modified as

$$A_P = \hat{I}_{in} T_s \sin(60^\circ - \theta_{SI}) k\left(\theta_{SI}, \theta_{SV}\right), \tag{3.30}$$

$$\text{where,} \quad k\left(\theta_{SI}, \theta_{SV}\right) = \left\{1 + \sqrt{3} m \sin(\theta_{SI} - 30^\circ)\cos(\theta_{SV} - 30^\circ)\right\}. \tag{3.31}$$

Since θ_{SI}, θ_{SV} each varies in the closed interval [0, 60°] and m varies within [0,1], it is derived that A_p reaches its maximum when

$$\theta_{SI} = 28^\circ, \;\; \theta_{SV} = 60^\circ \;\; \& \;\; m = 1. \tag{3.32}$$

So, $\hat{v}_{rL}$ is maximum at an instant when $\bar{\mathbf{I}}_{\mathbf{in}}$ is almost at the middle of sector II i.e. when $i_{fa} \approx i_{fb}$.

The values of θ_{SI}, θ_{SV} and m from (3.32) are substituted in (3.30) and (3.31) to evaluate the maximum value of A_p. Subsequently using this value in (3.22), maximum $\hat{v}_{rL}$ is derived as

$$\hat{v}_{rL,\,\max} \approx \frac{1}{2C_f}\hat{I}_{in} T_s. \tag{3.33}$$

From (2.27), the input current amplitude is

$$\hat{I}_{in} = \frac{\sqrt{3}}{2} m\, \hat{I}_o \cos(\varphi_{oL}). \tag{3.34}$$

Hence the maximum possible value of $\hat{I}_{in}$ is $(\sqrt{3}/2)\hat{I}_o$, which on substitution in (3.33) gives the maximum peak-peak magnitude of the ripple voltage as

$$\hat{v}_{rL,\,\max} \approx \frac{\sqrt{3}}{4C_f}\hat{I}_0 T_s. \tag{3.35}$$

Incorrectly sensing this ripple voltage polarity may lead to an incorrect commutation sequence as detailed below.

3.1.4.2 Error in Commutation

Figure 3.9 plots the ripple current and v_{rL}, in the neighbourhood of the instant when $\hat{v}_{rL}$ reaches $\hat{v}_{rL,\,\max}$. This is when commutation from phase-a to phase-b takes place. The figure depicts a situation where the capacitor voltages are sampled at a frequency $4f_s$. Let v_{rs} be the ripple component of the sensed voltage, which is subsequently available as the input signal for sampling. Noise filtering in the voltage transducer electronics and the anti-aliasing filters, prior to sampling, invariably cause delays in the sensed signal. The plot of v_{rs}, shown in the figure, is obtained after considering the most realistic structure of the measurement circuit and filter, viz. first-order, low-pass.

Fig. 3.9 Error in measuring v_{rL} around its maximum value

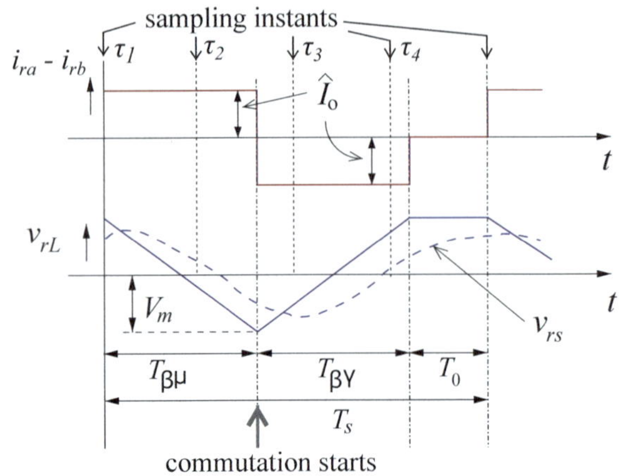

The sampling instants before commutation are shown in Fig. 3.9. Commutation starts at the instant indicated and the commutation sequence is based on the signals sampled just prior to it, i.e. at the instant marked as τ_2 in the figure. At this instant (τ_2), the sampled value of v_{rs} is positive while v_{rL} is negative. At the instant commutation starts, v_{rL} reaches its negative maximum value, V_m. Using (3.35),

$$V_m \approx -\frac{\sqrt{3}}{8C_f}\hat{I}_o T_s. \tag{3.36}$$

The 3 Ph waveforms, shown in Fig. 3.10, have been used to identify the different variables, based on their instantaneous phases at the start of commutation. At the instant when commutation is just about to begin, $\bar{\mathbf{I}}_{\mathbf{in}}$ is very close to the middle of sector II of input current hexagon. Therefore, the intersection of the vertical line L1 with the three waveforms, indicates the position of i_{fa}, i_{fb} and i_{fc} respectively at the beginning of commutation. At the same instant, the relative position of the fundamental components of input voltages depends on the IDF. Let, intersection of the line L2 with the waveforms indicate positions of v_{fa}, v_{fb} and v_{fc} respectively. So,

Fig. 3.10 Instantaneous positions of i_{fa}, i_{fb}, v_{fa} and v_{fb} at the beginning of commutation from phase-a to b

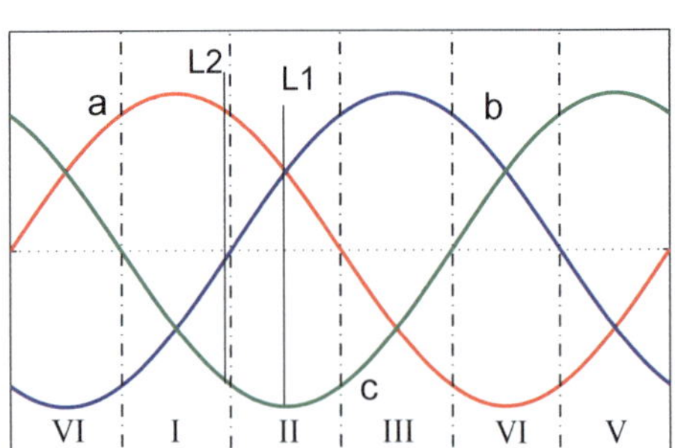

here the input currents lead the voltages, a situation that can arise in applications requiring input reactive power control [18, 19].

Reported voltage based commutation (VBC) strategies recommend identification of the two phases involved in 'critical' and 'uncritical' commutation, on basis of the input voltage magnitude [16]. From Fig. 3.10, since v_{fa} and v_{fb} have distinctly different instantaneous values, commutation between these two phases is to be interpreted as uncritical. For uncritical commutation, regular 2 or 4 step commutation methods are suggested [16, 17]. It is shown now that even in the presumable 'safe' situation, there is a strong possibility of wrong sequence selection. This can be avoided if the filter capacitor is appropriately designed, as explained below.

Instantaneous values of the fundamental components of the input capacitor voltages in phases-a and b, at this instant (along L2), are,

$$\left.\begin{aligned} v_{fa} &\approx \frac{\sqrt{3}}{2}\hat{V}_{cf} \\ v_{fb} &\approx 0 \end{aligned}\right\} \Rightarrow v_{fL} = \frac{\sqrt{3}}{2}\hat{V}_{cf}. \tag{3.37}$$

Since v_{fL} has a fairly high magnitude and a much lower frequency, the delay in sensing its polarity is much lesser than that of the ripple component. It would therefore be realistic to assume that its polarity is correctly detected. Also, since the sampled value of v_{rs} is positive, v_L is also interpreted to be positive. The limit within which this interpretation remains correct forms one basis for choosing the minimum value of the filter capacitor.

Using (3.36) and (3.37) gives v_L as

$$v_L = v_{fL} + v_{rL} = \frac{\sqrt{3}}{2}\left(\hat{V}_{cf} - \frac{\hat{I}_o T_s}{4C_f}\right). \tag{3.38}$$

Therefore, v_L becomes negative if

$$\hat{V}_{cf} < \frac{\hat{I}_o T_s}{4C_f} \tag{3.39}$$

and this can be avoided when

$$C_f \geq C_{f,\,\min}^{(2)} = \frac{\hat{I}_o T_s}{4\hat{V}_{cf}} \approx \frac{I_o T_s}{4V_s}. \tag{3.40}$$

If this condition (3.40) is violated, the commutation process from phase-a to b, initiated with an incorrect sequence corresponding to positive polarity of v_L, may lead to commutation failure. Figure 3.11 shows the condition that will arise in this situation of incorrect commutation sequence. Short circuit current i_{sc} flows over an interval T_{sc}, which is the sum of two turn-on and turn-off intervals of the IGBT. L_{st}

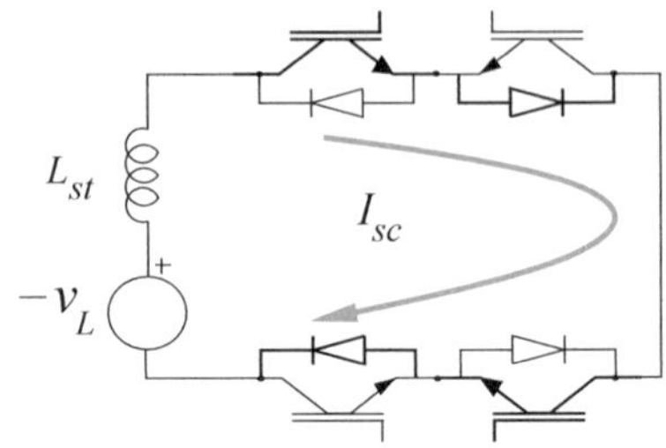

Fig. 3.11 Path of short circuit current

in Fig. 3.11 denotes the stray inductance of the circuit. Let v_D be the total forward voltage drop of all the devices in the short-circuit path comprising 2 IGBTs and 2 diodes. If $\hat{I}_D$ is the maximum current rating of the devices, the criterion to ensure device safety is obtained as

$$I_{sc,\,\max} = \frac{1}{L_{st}} \int_0^{T_{sc}} (-v_L - v_D)\, dt \le \hat{I}_D. \tag{3.41}$$

Ignoring the non-linearity of the device parameters and using (3.38) in (3.41), another criterion for the lower limit of C_f is obtained as

$$C_{f,\,\min}^{(3)} = \frac{1}{4} \frac{\hat{I}_o T_s}{\left\{\hat{V}_{cf} + 1.15\left(v_D + \frac{L_{st}\hat{I}_D}{T_{sc}}\right)\right\}}. \tag{3.42}$$

which is obviously hardware specific. $C_{f,\,\min}^{(2)}$ may be used as a preliminary check before applying the last criterion, which is dependent on the hardware details.

Table 3.8 shows $C_{f,\,\min}^{(2)}$ and $C_{f,\,\min}^{(3)}$, calculated using tabulated device ratings and measured L_{st}. $C_{f,\,\min}^{(2)}$ represents the minimum value of C_f required for safe commutation for any operating condition. Hence, $C_{f,\,\min}^{(2)}$ must be preferred over $C_{f,\,\min}^{(3)}$. The final choice of $C_{f,\,\min}$ is chosen as

$$C_{f,\,\min} = \sup\{C_{f,\,\min}^{(1)}, C_{f,\,\min}^{(2)}, C_{f,\,\min}^{(3)}\}. \tag{3.43}$$

Table 3.8 $C_{f,\,\min}$

$C_{f,\,\min}^{(2)}$	$C_{f,\,\min}^{(3)}$	T_{sc}	L_{st}	$\hat{I}_D$	v_D
1.04 μF	0.97 μF	2 μs	260 nH	80 A	10.1 V

Adhering to all described limits, chosen Q and f_c, the parameters selected for input filter are provided in Table 3.10.

So far, L_s has been neglected in the analysis. Its impact on filter performance is discussed subsequently.

3.1.5 Effect of L_s

With the source inductance, L_s, considered in the circuit, forward gain of the input filter becomes

$$G_{fv}(s) = \frac{s\frac{L_f}{R_d} + 1}{s^3\frac{L_sL_fC_f}{R_d} + s^2C_f(L_s + L_f) + s\frac{L_f}{R_d} + 1}. \tag{3.44}$$

In its denominator polynomial,

$$D(G_{fv}) = s^3 \underbrace{\frac{L_sL_fC_f}{R_d}}_{P_1} + s^2 \underbrace{(L_s + L_f)C_f}_{P_2} + s \underbrace{\frac{L_f}{R_d}}_{P_3} + 1, \tag{3.45}$$

finding accurate solutions for the roots can be quite complex. Consequently it becomes very difficult to understand the effects of incorporating L_s. This task becomes feasible with a reasonably accurate lower order approximation of $D(G_{fv})$. A second order approximation is considered based on the assumption that $P_2 \gg P_1$. The basis of this assumption is that in all practical installations L_s would be in the order of mH and R_d in tens of Ωs. So $P1 \sim P2 * (L_s/R_d)$ is likely to be quite small compared to $P2$. This leads to the following approximated expression,

$$G_{fv}(s) \approx \frac{s\frac{L_f}{R_d} + 1}{\left(s\frac{L_f}{R_d}\frac{n}{n+1} + 1\right)\left(\frac{s^2}{\omega_{cc}^2} + \frac{s}{\omega_{cc}Q_c} + 1\right)} \tag{3.46}$$

where

$$\omega_{cc} \approx \frac{1}{\sqrt{(n+1)L_fC_f}}, \quad Q \approx R_d\left(\sqrt{\frac{C_f}{L_f}}\right)(1+n)^{1.5}, \quad n = \frac{L_s}{L_f}. \tag{3.47}$$

Figure 3.12 shows the gain plot for different values of L_s. Input filter parameters are used from Table 3.10.

So with increasing L_s, Q increases and ω_{cc} falls which leads to increase in the gain to low order harmonics. However the losses due to the switching ripple components in i_{in} would decrease since a larger impedance appears across C_f which reduces the ripple current flowing into the source, v_s. Hence a knowledge of L_s is necessary before finalizing on the input filter parameters. With n as defined is (3.47), the nat-

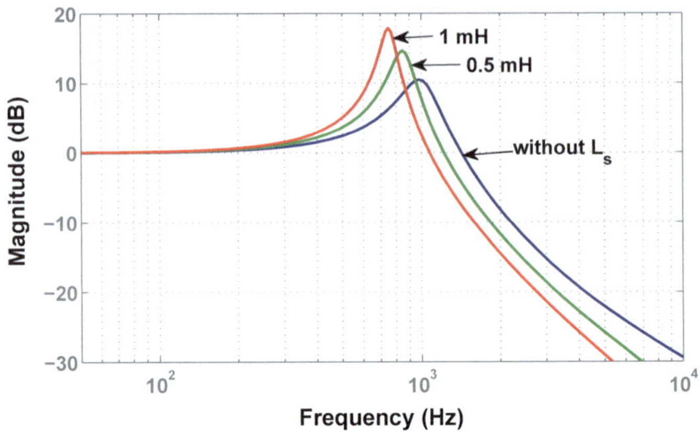

Fig. 3.12 Magnitude plot of $G_{fv}(s)$ for L_s of 0.5 and 1 mH

ural frequency of the filter is not significantly affected, i.e. $\omega_c \approx \omega_{cc}$, if $n < 0.5$, with minimal impact on filter performance. A high n, however, further complicates the controller design, apart from deteriorating the harmonic performance as discussed subsequently.

3.1.6 Design Modifications for a Non-minimum Phase Plant

Recalling the condition of minimum phase plant from (2.106),

$$P_{in} < \frac{\frac{3}{2}\hat{V}_{cf}^2 \cos^2 \varphi_i}{|Z_{so11}(j\omega_M)|} = P_{cr}. \tag{3.48}$$

With reference to (3.48),

$$|Z_{so11}(j\omega_M)| \approx ||Z_{so11}(j\omega)||_\infty \leq ||Z_{so}(j\omega)||_\infty. \tag{3.49}$$

Assuming underdamped second-order filter and ideal passive elements

$$||Z_{so}(j\omega)||_\infty = |Z_{so}(j\omega_c)| \approx R_d \left(1 + \frac{L_s}{L_f}\right)^2. \tag{3.50}$$

So as L_s increases, there is an increasing chance of the plant transforming itself from a minimum to non-minimum phase system. With non-minimum plant zeros, the closed-loop system is bound to be unstable as the algebraic gain increases. This places severe constraints on the close-loop bandwidth and thus certainly makes controller design extremely difficult since as many as 4 RHZs emerge if (3.48) is violated. In such a situation, either the dynamic specifications have to be made less stringent or the input filter requires redesign. As a way out, choice of a lower R_d or higher L_f can be opted for to ensure that (3.48) is satisfied and the plant remains minimum phase till its declared rated operation. Both of these options degrade steady state performance in different ways. Increasing L_f causes greater voltage drop across it, which reduces the maximum output voltage that can be synthesized. The other option of lowering R_d increases losses. So, for a weak grid, characterized by high L_s, a trade-off is necessary between dynamic performance and either filter losses or full-load regulation.

3.2 Output Filter

Denoting the natural frequency and quality factor of the output filter shown in Fig. 3.13 as ω_{co} and Q_o respectively, the forward gain defined in (2.50) is obtained as

$$G_{vof}(s) = \frac{v_{co}(s)}{v_{on}(s)} = \frac{1}{(s/\omega_{co})^2 + (s/Q_o\omega_{co}) + 1}. \tag{3.51}$$

Q_o can be controlled by using virtual damping element which would be discussed in a later chapter. Selection of Q_o is carried out using the same approach as done during input filter design. Criteria defined in **Spec.1–Spec.4**, are directly followed for selecting $f_{co,\ max}$, $f_{co,\ min}$, $C_{o,\ max}$ and $L_{o,\ max}$. The most significant harmonic at the output side is assumed same as that of the input side, since literature on exhaustive harmonic analysis of the modulation scheme is still unavailable. Hence the A_{vh}, listed in Table 3.2, is used here. Using the nominal values in Table 3.1, the range of natural frequency and maximum limits for the filter elements are listed in Table 3.9.

Recall that the ripple current rating of the input filter capacitor was decided by the output current. Interestingly, appropriate sizing of the output filter inductor, L_o, permits use of capacitors with lower ripple current rating. This is detailed as follows.

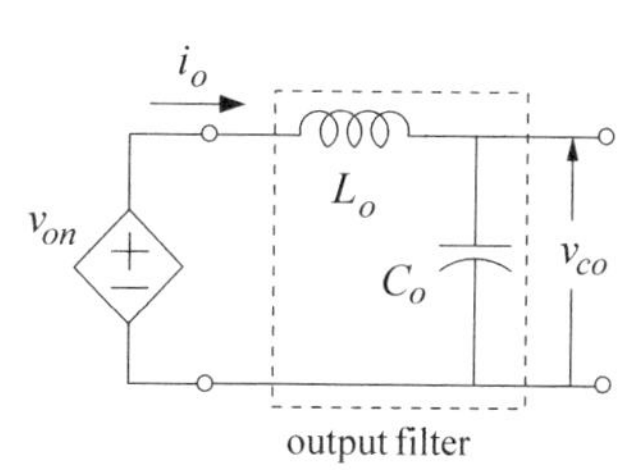

Fig. 3.13 Output filter

Table 3.9 Boundary values for parameters of output filter

$f_{co,\ max}$	$f_{co,\ min}$	$L_{o,\ max}$	$C_{o,\ max}$	$L_{o,\ min}$
2.2 kHz	777 Hz	2 mH	26.5 μF	1.6 mH

3.2.1 Lower Limit of L_o

Output admittance, or the reciprocal of the output impedance, is obtained as

$$Y_{mo}(s) = \frac{i_o(s)}{v_{on}(s)} = \frac{1}{R_o}\frac{1 + s\,(Q_o/\omega_{co})}{(s/\omega_{co})^2 + (s/Q_o\omega_{co}) + 1}. \tag{3.52}$$

For Q_o greater than 1,

$$|Y_{mo}(j\omega_s)| \approx \frac{1}{\omega_s L_o}. \tag{3.53}$$

In SVM [25], at any instant, 3 output phases are connected to any two input phases. Hence, from Fig. 2.1, for output phase A,

$$\hat{v}_{on} = \frac{2}{3}\hat{v}_{ab} = \left(\frac{2}{3}\right)\sqrt{2}(\sqrt{3}V_s) = 1.63V_s. \tag{3.54}$$

As in the case of the input filter, assuming the maximum possible amplitude of the switching ripple component in the output voltage to be equal to that of the fundamental, the maximum ripple current is easily calculated using (3.53) and (3.54) as,

$$\hat{I}_{ro} \leq \frac{1}{\sqrt{2}}\left(\frac{\hat{v}_{on}}{\omega_s L_{o,\min}}\right). \tag{3.55}$$

Note that $\hat{I}_{ro}$ also demarcates the minimum ripple current rating ($\hat{I}_{r,\ \text{rated}}$) of C_o. Since this value is directly decided by $L_{o,\min}$, it allows arbitrary choice of the ripple current rating of the filter capacitor. Hence,

Spec.6 $L_{o,\ \min} = \dfrac{1.63V_{s,\ \text{rated}}}{\omega_s\hat{I}_{r,\ \text{rated}}}.$

Table 3.9 lists the described limits for the output filter parameters calculated using the previous considerations.

Table 3.10 lists a specific solution set, which was used for the experimental investigation. Of course, this set completely satisfies all the design rules laid down so far and the verification could familiarize the reader with the design process.

Table 3.10 Input and output filter parameters

Input filter			Output filter	
C_f	L_f	R_d	C_o	L_o
20 μF	1.26 mH	25 Ω	20 μF	2 mH

3.3 Selection of Parameter Values

A number of design guidelines have emerged from the last two sections. Solution sets which ensure compliance with different criteria has been identified for both input and output filters. These guidelines are summarized here to aid a structured design process. Pre-requisites for the design are the numerical values of the nominal power, fundamental frequency of the grid where the MC is to be deployed, switching frequency and percentages of different harmonic in the source voltage. Filter parameters are then selected by following these steps.

Step 1. Select the quality factor Q ($=Q^{(1)}$) from the normalized form of the input filter forward gain. Use the design procedure detailed before to conform to **Spec.1** and **Spec.2**. Select the the optimum value of Q ($Q^{(1)}$), as discussed in Sect. 3.1.3. This is the finalized value for design.

Step 2. Use the normalized forward gain plot again to decide first the range of f_c. You will need to also use the value of $Q^{(1)}$, that was decided in Step 1, along with the numerical values of **Spec.1** and **Spec.2**. Then choose a value within this range.

Step 3. Then compute the upper limits $L^{(1)}_{f,\,\max}$ and $C^{(1)}_{f,\,\max}$. For this you need to use **Spec.3** and **Spec.4** respectively. Use these values along with the selected f_c and calculate the corresponding $C^{(1)}_{f,\,\min}$ and $L^{(1)}_{f,\,\min}$.

Step 4. Now you should find the first allowable set of R_d, which has to conform to **Spec.5**. Then compute the second set of R_d based on **Spec.3** and **Spec.4**, as per the discussions in Sect. 3.1.3. The intersection of these two sets form the final allowable range of R_d. If they do not intersect, then either one or both of the sets has to be expanded by relaxing the limits of either **Spec.5** or any one out of **Spec.3** and **Spec.4** till the intersection set ceases to be a nullity. Relaxing **Spec.5** entails higher losses. Be aware that relaxing limits of **Spec.3** implies poorer regulation and those of **Spec.4** leads to higher reactive loading. You have to choose which limit to relax based on what the target application demands. For example, if the application requires operating the MC at a high modulation index then **Spec.3** should be kept tight and any relaxation would seriously hamper performance. Conversely, for operation with low modulation index, relaxing this limit is preferable over the other two. So, in case the intersection set is a nullity, this step would require several design iterations.

Step 5. Using (3.40), compute the value of $C^{(2)}_{f,\,\min}$ to ensure reliable voltage commutation. If you have reasons to doubt the linearity of the voltage sensor, like offsets, or any other aspect of the voltage measurement circuit, then also compute $C^{(3)}_{f,\,\min}$. But for this, you will need to have an idea of the parasitic inductance in the

Table 3.11 Input and output filter parameters

Parameter	Value	Sp. volume (cm^3/kW)	Sp. weight (kg/kW)
C_f	20 μF	19.16	0.0316
L_f	1.26 mH	101	0.38
R_d	25 Ω		
C_o	20 μF	0.93	0.00125
L_o	2 mH	91	0.38

switch layout of the MC. The absolute lower limit for C_f, i.e. $C_{f,\text{ min}}$, is decided according to (3.43).

Step 6. Select L_f from the interval $[L^{(1)}_{f,\text{ min}}, L^{(1)}_{f,\text{ max}}]$. Also choose a value for C_f from the interval $[C_{f,\text{ min}}, C_{f,\text{ max}}]$, as discussed in *Step 5*. Also, finalize the value of R_d from the range calculated in *Step 4*.

Step 7. This is a critical step where the finalized input filter parameters are substituted in (3.50) and then in (3.48) to detect whether the design choice results in a minimum phase plant. If it does not, then a preliminary controller design is required to find out the minimum value of the algebraic gain of the loop transfer function which satisfies the dynamic response requirements. With this gain, location of the close-loop poles needs to be determined from a root-locus analysis. No right-half poles indicate stable operation but make sure all parameter variations are considered. A safer option is to opt for a minimum phase plant in which case the limits for either full-load regulation (**Spec.3**) or filter loss (**Spec.5**) has to be relaxed. Input filter parameters have to be redesigned repeating *Steps 3–6* and cross-checked in *Step 7*. This would complete the input filter design.

Step 8. Select the parameters of the output filter, i.e. compute the values of Q_o, f_{co}, $C_{o,\text{ max}}$ and $L_{o,\text{ max}}$ following the same steps as those for the input filter.

Step 9. Select $L_{o,\text{ min}}$ so as to conform to **Spec.6**. You have an extra flexibility in being able to choose the ripple current rating of C_o when you apply **Spec.6**. Then choose the values of L_o and C_o. This completes the design process.

For the experimental system, the filter parameters are designed by following the steps described, with the boundary values listed in Tables 3.2, 3.3, 3.6, 3.8 and 3.9. Values chosen for the filter parameters in the experimental system are shown in Table 3.11. The specific volume (cm^3/ kW) and weight (kg/kW) of the inductors and capacitors used in experiment have also been listed in Table 3.11. Experimental results are discussed in the following section.

3.4 Results and Discussion

Experimental validation was carried out on a 6 kVA MC prototype. Switches used were IGBT based 4-quadrant switch modules from SEMIKRON. The entire control logic was coded on an FPGA platform configured around the ALTERA

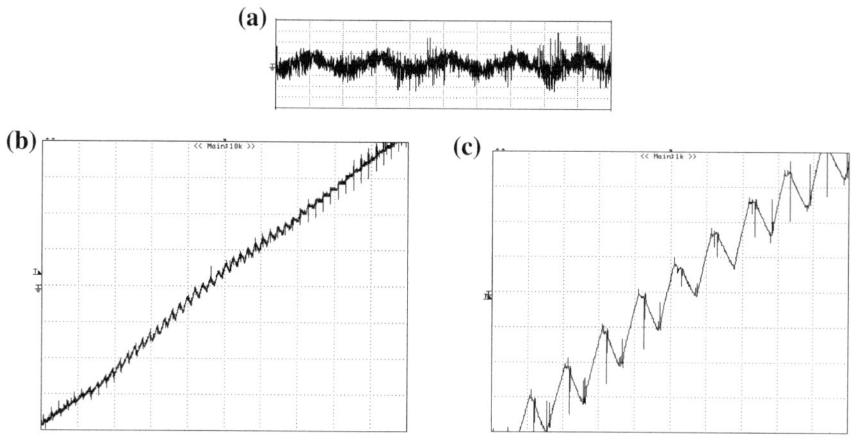

Fig. 3.14 **a** Voltage across R_d: 10 V/div, time: 10 ms/div. **b** v_L: 100 V/div, time: 500 μs/div. **c** v_L: 20 V/div, time: 100 μs/div

EP1C12Q240C8 processor. The control code was run with a sampling frequency of 20 kHz. For commutation, v_{cf} was measured and subsequently sampled at 40 kHz.

3.4.1 Open Loop Experimental Results

The experimental MC was operated at $m = 1$ and unity IDF, with fundamental output current, $i_o = 8.2$ A. Figures 3.14, 3.15 and 3.16 shows the experimentally obtained waveforms. Measured data are listed in Table 3.12.

Figure 3.14a plots the voltage across the damping resistor, R_d. From the experimental data, its r.m.s value was calculated which was subsequently used to compute the power loss, P_R. This experimentally determined value was 2 W, which is much lower than 1% of the input power. Analysis in Sect. 3.1.3 using fundamental v_s and i_o leads to a theoretical value of 2.1 W, which is very close to that experimentally obtained.

Referring to Sect. 3.1.4 and particularly (3.32), unity IDF operation implies that the maximum of the input capacitor ripple voltage, $\hat{v}_{rL,\,\max}$, is expected to appear almost at the ZC of the input line-line voltage v_L. Experimental corroboration is provided in Fig. 3.14b, which plots v_L over an interval of 5 ms around its ZC. A zoomed view of the same waveform is presented in Fig. 3.14c, for an interval of 1 ms. Note that maximum ripple occurs around the ZC zone, as predicted by the analysis. Impedance measurement of the input filter capacitors was conducted with a precision RCL meter and the measured values are listed in Table 3.12, for a 10 kHz test frequency. Values of C_f from Table 3.12 are used to calculate v_{ra} and v_{rb} at the maximum ripple condition as detailed in Sect. 3.1.4 and the theoretically calculated value of $\hat{v}_{rL,\,\max}$ magnitude is found to be 30.8 V. For validation, the switching ripple

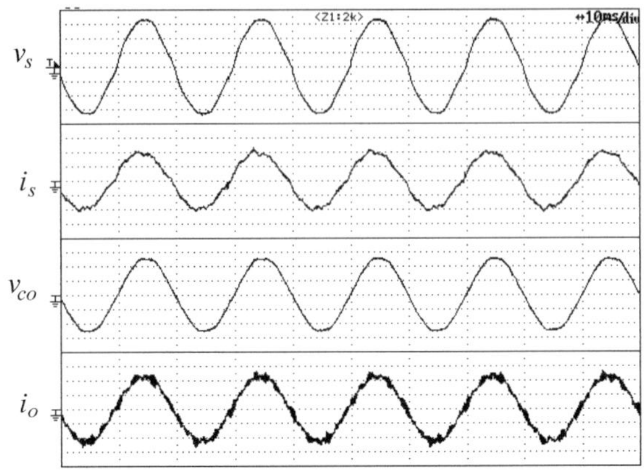

Fig. 3.15 Open loop experimental results. v_s: 100 V/div, i_s: 5 A/div, v_{co}: 100 V/div, i_o: 5 A/div. Time: 10 ms/div

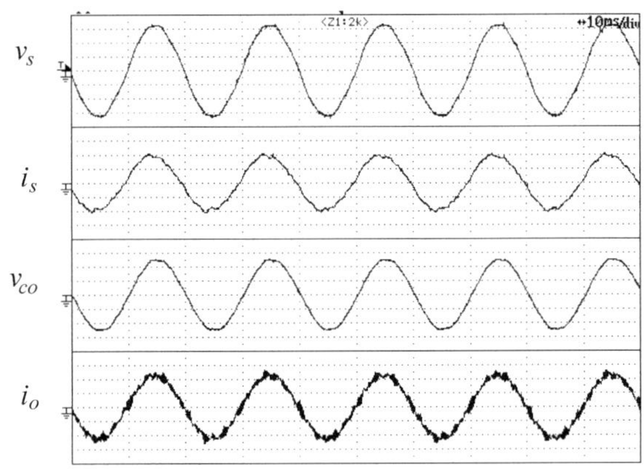

Fig. 3.16 Open loop results with a L_s of 1mH. v_s: 100 V/div, i_s: 5 A/div, v_{co}: 100 V/div, i_o: 5 A/div. Time: 10 ms/div

Table 3.12 Measured values

Input power	C_f (phase-a)	C_f (phase-b)	$\hat{v}_{rL,\,\max}$	IDF
4.9 kW	17.9 μF	14.5 μF	31 V	0.982

component is numerically extracted from the measured capacitor voltage data. This is done off-line by using a low-pass filtering on the measured data, with the filter corner less than a decade below the switching frequency, and subsequent subtraction of the filter output from its input signal. The maximum value of the experimentally obtained input capacitor ripple voltage is listed in Table 3.12. Again, a very close agreement is evident between the analytical and experimental observations. Experimental data also revealed an IDF of 0.982 as listed in Table 3.12.

Table 3.13 lists the harmonic components in phase-a of the supply voltage v_s, which is seen to contain significant low order harmonics. From the harmonics of v_{co} listed, it is evident that the gain criterion adopted in Sects. 3.1 and 3.2 is effective in minimal transmission of low order harmonics from v_s to v_{co}. Since i_s is a function of

Table 3.13 Harmonics in one phase of v_s, i_s and v_{co}

Harmonic order	v_s (%)	i_s (%)	v_{co} (%)
3rd	1.1	2.65	0.3
5th	2.64	2.16	1
7th	2	4	1
9th	0.54	2	0.6
11th	0.23	0.2	0.2
THD	3.6	5.9	1.67

Table 3.14 Harmonics v_s, i_s and v_{co} with $L_s = 1$ mH

Harmonic order	v_s (%)	i_s (%)	v_{co} (%)
3rd	1.8	3.2	0.6
5th	1.9	1	1
7th	1	2.2	0.9
9th	0.9	2.6	0.6
11th	0.1	0.65	0.15
15th	0.1	1	0.05
17th	0.1	0.9	0.1
THD	3.1	5.4	1.8

both v_s and i_o, it has a higher harmonic content. Switching ripple in i_s was measured to be slightly less than 1% of the fundamental.

3.4.1.1 Performance with L_s Included

To validate the analytical results with source inductance, L_s, a 1 mH inductor was externally added. Figure 3.16 shows the steady-state waveforms, at the same power level, as that of the earlier set of experimental results. With this inclusion, using (3.47), ω_c reduces from 1 kHz to 746 Hz and Q increases from 3.1 to 7.6. Measured harmonic content of v_s, i_s and v_{co} are listed in Table 3.14. Owing to the proximity of input and output corner frequencies (746 and 796 Hz) an increase in harmonic content of i_s close to these frequencies is observed. When this second set of experiments were performed, THD of v_s was slightly better than the previous situation. However due to proximity of the input and output filter corners, harmonics of i_s around the output corner were observed to have increased which marginally increased the THD of v_{co}. The measured value of IDF was found to be 0.984.

3.5 Concluding Remarks

This chapter provided a comprehensive design approach for the input and output filters for the MC. Aspects of filter design like attenuation of switching frequency components, regulation, losses in filter and also MC-specific issue like the damping resistor for input filter were addressed. A worst case analysis of the associated losses is provided. The design process referred to real-life numbers derived from an experimental system. Experimental results are presented and discussed which validate the analytical limits. Thereafter, the condition for the maximum input voltage ripple has been evaluated. Experimental results demonstrate a very close agreement with the analysis. Minimum input filter capacitor required for reliable voltage based commutation based on the maximum error in sensing the ripple voltage has been derived. The effect of the grid inductance has been discussed where it is shown that a compromise might become necessary between filter and controller design requirements. The major differences in output ripple filter design are also presented, the rest being similar to the input filter design approach has been addressed in summary to avoid repetition. Experimental waveforms are provided to demonstrate filter performance with the designed parameters, in accordance with the specifications.

Chapter 4
Controller Design for Regulated Voltage Supply

Controller design for applications requiring regulated 3 phase sinusoidal voltages is discussed and relevant experimental results are presented in this chapter. The controller design is based on the dynamic model described in Chap. 2. The model represented a two input–two output system and thus the controller cannot be designed by directly applying single variable based design methods. Also, the plant has been shown to contain zeros which migrate towards the right half plane as the throughput power increases. In frequency domain based design methods for non-minimum phase plants it is imperative for internal stability that the right plane zero/pole must not be cancelled by the controller poles/zeros. Keeping this in mind, the design problem has been reduced to a single variable control design problem without violating internal stability.

In the initial sections of this chapter, transformation of the multi variable control problem into a single input single output (SISO) control problem has been described. The design and realization of the controller is subsequently discussed. Experimental results using the lab prototype MC is presented to validate the analysis. The results include operation beyond the critical power (P_{cr}) region where the non-minimum phase zeros appear.

4.1 Specifications and Assumptions

Controller design is initiated with appropriate specifications for the close-loop system dynamics. Typical performance parameters chosen are settling time and phase margin as shown in Table 4.1. This particular choice of settling time and phase margin ensures one-cycle response for synchronous applications with 50 Hz systems depicting first order closed loop characteristic. Also, for the consequent closed-loop bandwidth of ω_{BW} (= 200 rad/s), the choice of sampling frequency ω_{sp} (= 20 kHz) minimizes the phase delay due to discretization $\left(= \tan^{-1}\left[\omega_{BW}/\omega_{sp}\right]\right)$ at ω_{BW} to a negligible value. Under similar consideration, since the sensors used in the test set-up have a much

A. Dasgupta and P. Sensarma, *Design and Control of Matrix Converters*,
Energy Systems in Electrical Engineering, DOI 10.1007/978-981-10-3831-0_4

Table 4.1 Dynamic specifications

Settling time	Phase margin
20 ms	90°

higher bandwidth than ω_{BW}, their delays have not been considered in the feedback path.

In this chapter, the emphasis is to demonstrate stable operation beyond P_{cr}. Therefore moderate and heavy resistive loads have been considered. Regulation at low load or high reactive loads would arise during discussion on voltage sag mitigation, where, the necessary modifications in control would be discussed. Design of voltage controller is detailed in the following subsections.

4.2 SISO Based Control

The input-output equations for the linearized model, derived in Chap. 2, are

$$\tilde{\mathbf{v}}_{\mathbf{co}}(s) = \begin{bmatrix} \tilde{v}_{cod}(s) \\ \tilde{v}_{coq}(s) \end{bmatrix} = \mathbf{G_{vof}}(s)\mathbf{G_I}^{-1}(s)\mathbf{G_a}(s) \begin{bmatrix} \tilde{m}_d(s) \\ \tilde{m}_q(s) \end{bmatrix} = \mathbf{G_c}(s) \begin{bmatrix} \tilde{m}_d(s) \\ \tilde{m}_q(s) \end{bmatrix}. \quad (4.1)$$

If this problem has to be reduced to a SISO problem, then any coupling factor between the d-axis and q-axis variables has to be removed first. This decoupling should be done without any right plane pole-zero cancellation so that the internal stability of the system is not compromised.

In the the plant represented in (4.1) by the transfer function matrix $\mathbf{G_c}(s)$, the only possible source of any non-minimum phase component is in form of the right half plane zeros (RHZs) of $\mathbf{G_a}(s)$. Consequently, removal of coupling components of $\mathbf{G_a}(s)$ should not occur at the expense of cancellation of any of its RHZs. Since the rest of the plant transfer matrix represented by $\mathbf{G_{vof}}(s)\mathbf{G_I}^{-1}(s)$ contains only minimum phase poles and zeros, decoupling for this part can be achieved by simply multiplying with inverse transfer function matrix. The following discussion starts with decoupling of $\mathbf{G_a}(s)$ and subsequent reduction to SISO control problem.

It was shown in Chap. 2 that $\mathbf{G_a}(s)$, derived as

$$\mathbf{G_a}(s) = \begin{bmatrix} a_V - M_d Z_{so11}(s) I_{od} & -M_d Z_{so11}(s) I_{oq} \\ -M_q Z_{so11}(s) I_{od} & a_V - M_q Z_{so11}(s) I_{oq} \end{bmatrix}, \quad (4.2)$$

can also be represented in the following form

$$\mathbf{G_a}(s) = \frac{1}{d(s)} \underbrace{\begin{bmatrix} a_{11}(s) \; a_{12}(s) \\ a_{21}(s) \; a_{22}(s) \end{bmatrix}}_{[\mathbf{N}(s)]}, \quad (4.3)$$

where, d(s) is the common denominator polynomial and $\mathbf{N}(s)$ is a matrix made up of polynomials. The poles and zeros of $\mathbf{G_a}(s)$ are a subset of the roots of the following equations

$$d(s) = 0 \quad \& \quad det(\mathbf{N}(s)) = 0 \Rightarrow a_{11}(s)a_{22}(s) - a_{12}(s)a_{21}(s) = 0, \tag{4.4}$$

respectively. A matrix can be thought of as [45]

$$\mathbf{N}_{decp}(s) = \begin{bmatrix} a_{22}(s) & -a_{12}(s) \\ -a_{21}(s) & a_{11}(s) \end{bmatrix} \tag{4.5}$$

such that

$$\mathbf{N}(s)\mathbf{N}_{decp}(s) = \begin{bmatrix} det(\mathbf{N}(s)) & 0 \\ 0 & det(\mathbf{N}(s)) \end{bmatrix} = det(\mathbf{N}(s))^2 \begin{bmatrix} 1 & 0 \\ 0 & 1 \end{bmatrix} = det(\mathbf{N}(s))\mathbf{I}_2. \tag{4.6}$$

There are two important things to note here. First, both $\mathbf{N}(s)$ and $\mathbf{N}_{decp}(s)$ has zeros in the same locations. Second, the product of both contains a scalar transfer function which also has its zeros at the same location as $\mathbf{N}(s)$ and $\mathbf{N}_{decp}(s)$. Moreover, this scalar transfer function is obtained without canceling any zeros of the two matrices. Therefore this is an useful tool to decouple without violating internal stability.

Denoting the voltage error as $\tilde{\mathbf{e}}_\mathbf{v}(s)$, the feedback control design problem can be represented as

$$\tilde{\mathbf{v}}_\mathbf{co}(s) = \mathbf{G_c}(s) \begin{bmatrix} \tilde{m}_d(s) \\ \tilde{m}_q(s) \end{bmatrix} = \mathbf{G_c}(s) \underbrace{\mathbf{N_a}(s)\mathbf{N_{mp}}(s)}_{\mathbf{N_{ff}}(s)} \underbrace{H_v(s)\tilde{\mathbf{e}}_\mathbf{v}(s)}_{\tilde{\mathbf{u}}(s)}, \tag{4.7}$$

where, $\mathbf{N_a}(s)\mathbf{N_{mp}}(s)H_v(s)$ or $\mathbf{N_{ff}}(s)$ are the feedforward compensators meant to decouple the plant $\mathbf{G_c}(s)$. $\tilde{\mathbf{u}}(s)$ is the output of the controller $H_v(s)$. The design task can be viewed as the following set of tasks in a sequential manner

- deriving decoupling matrix $\mathbf{N_a}(s)$ for the non-minimum phase part i.e. $\mathbf{G_a}(s)$,
- deriving decoupling matrix $\mathbf{N_{mp}}(s)$ for the minimum phase part,
- finally, designing the controller $H_v(s)$ based on the decoupled scalar plant transfer function.

Using (4.2) to (4.5), $\mathbf{N_a}(s)$ can be derived as

$$\mathbf{N_a}(s) = \frac{1}{{a_V}^2} \begin{bmatrix} a_V - M_q Z_{so11}(s) I_{oq} & M_d Z_{so11}(s) I_{oq} \\ M_q Z_{so11}(s) I_{od} & a_V - M_d Z_{so11}(s) I_{od} \end{bmatrix}, \tag{4.8}$$

such that

$$\mathbf{G_a}(s)\mathbf{N_a}(s) = \left(1 - \frac{P_{in}}{\frac{3}{2}\hat{V}_{cf}^2 \cos^2 \varphi_i} Z_{so11}\right) \begin{bmatrix} 1 & 0 \\ 0 & 1 \end{bmatrix} = M_a(s)\mathbf{I}_2. \tag{4.9}$$

Using (4.9) in (4.7), the design problem gets modified to

$$\tilde{\mathbf{v}}_{\mathbf{co}}(s) = M_a(s)\mathbf{G}_{\mathbf{vof}}(s)\mathbf{G}_{\mathbf{I}}^{-1}(s)\mathbf{N}_{\mathbf{mp}}(s)H_v(s)\tilde{\mathbf{e}}_{\mathbf{v}}(s). \tag{4.10}$$

It is important to reiterate here that both $M_a(s)$ and $\mathbf{N}_{\mathbf{a}}^{-1}(s)$ contains the same set of zeros as $\mathbf{G}_{\mathbf{a}}(s)$. Moreover, the poles of $M_a(s)$ and common denominator of $\mathbf{N}_{\mathbf{a}}^{-1}(s)$ are same as that of $Z_{so11}(s)$, all of which lie in left half of s plane. Hence, all the possible non-minimum phase zeros of the plant are retained in $M_a(s)$. Therefore, internal stability of the system is not violated by this decoupling process.

Since, $\mathbf{G}_{\mathbf{vof}}(s)\mathbf{G}_{\mathbf{I}}^{-1}(s)$ represents the minimum phase part of $\mathbf{G}_{\mathbf{c}}(s)$, these can be compensated by simply choosing $\mathbf{N}_{\mathbf{mp}}(s)$ as

$$\mathbf{N}_{\mathbf{mp}}(s) = \mathbf{G}_{\mathbf{I}}(s)\mathbf{G}_{\mathbf{vof}}^{-1}(s). \tag{4.11}$$

Therefore using (4.10) and (4.11), the output $\tilde{\mathbf{v}}_{\mathbf{co}}(s)$ and the output of controller $H_v(s)$ i.e. $\tilde{\mathbf{u}}(s)$ are now related by a reduced order plant with scalar transfer function $M_a(s)$ as

$$\tilde{\mathbf{v}}_{\mathbf{co}}(s) = M_a(s)H_v(s)\tilde{\mathbf{e}}_{\mathbf{v}}(s) = M_a(s)\tilde{\mathbf{u}}(s). \tag{4.12}$$

This simplifies problem to a single variable controller design, where a $H_v(s)$ can be designed using SISO based methods. The process described from (4.9) to (4.12) is shown in Fig. 4.1.

The control equation can now be described as

$$\tilde{\mathbf{v}}_{\mathbf{co}}(s) = M_a(s)H_v(s)\tilde{\mathbf{e}}_{\mathbf{v}}(s) = M_a(s)H_v(s)\left[\tilde{\mathbf{v}}_{\mathbf{co}}^*(s) - \tilde{\mathbf{v}}_{\mathbf{co}}(s)\right]. \tag{4.13}$$

The overall compensator which has to be realized is

$$\mathbf{N}_{\mathbf{a}}(s)\mathbf{G}_{\mathbf{I}}(s)\mathbf{G}_{\mathbf{vof}}^{-1}(s)H_v(s) = \mathbf{N}_{\mathbf{ff}}(s)H_v(s). \tag{4.14}$$

Realization of the compensator and controller design are discussed subsequently.

4.2.1 $N_{ff}(s)$ and $H_v(s)$

From (4.14), it is evident that the order of $\mathbf{N}_{\mathbf{ff}}(s)$ is identical to $\mathbf{G}_{\mathbf{c}}(s)$. Hence realizing $\mathbf{N}_{\mathbf{ff}}(s)$ would lead to a very high order compensator (10 in the present case study). This is not necessary, as, in any controller design based on frequency response, the loop gain is always kept high in the low frequency region and low in the high frequency region. The dynamic response is dictated by the shape of the loop gain around ω_{BW}. So it is at ω_{BW}, where the plant has to be sufficiently well modelled.

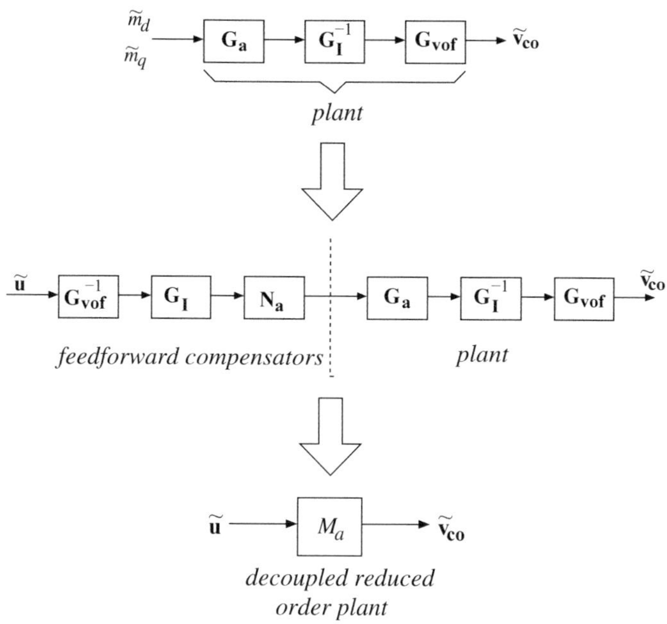

Fig. 4.1 Achieving a reduced order scalar plant $M_a(s)$

Adherence to the filter design rules described in Chap. 3, requires that both input and output filters should be designed to have their corner frequencies higher than the 7th harmonic frequency. So for the specified ω_{BW} of 200 rad/s, this means that the filter corners are at least a decade higher than ω_{BW}. Moreover, inclusion of $Z_{so11}(s)$ in the plant $M_a(s)$ means the resonant peak at input filter corner can be restricted by appropriately designing $H_v(s)$. High resistive load consideration implies that the output filter is reasonably well damped. Under these circumstances, chances of any unrealized higher frequency component of $\mathbf{N_{ff}}(s)$ adversely affecting the overall dynamics do not exist. Therefore, it is sufficient to realize a low frequency approximation of $\mathbf{N_{ff}}(s)$ up to ω_{BW}.

Therefore, denoting the low frequency approximation of $\mathbf{N_{ff}}(s)$ as $\mathbf{A_N}(s)$, the design objective is to ensure

$$\mathbf{A_N}(s) \approx \mathbf{N_{ff}}(s) \quad \text{up to } \omega_{BW}. \tag{4.15}$$

As the first iteration, the DC gain matrix $\mathbf{N_{ff}}(s)|_{s=0}$, is considered to be the low frequency approximation, $\mathbf{A_N}(s)$. The DC gain of $\mathbf{N_{ff}}(s)$ is

$$\mathbf{N_{ff}}(s)|_{s=0} \approx \frac{1}{a_V}\begin{bmatrix} 1 & \omega_T/(\omega_{co}Q_o) \\ -\omega_T/(\omega_{co}Q_o) & 1 \end{bmatrix}. \tag{4.16}$$

Table 4.2 System parameters

Parameters		Value
Base frequency		50 Hz
Nominal Supply voltage		240 V
Switching frequency		10 kHz
Sampling frequency		20 kHz
Source inductance	L_s	3 mH
Input filter inductance	L_f	1.26 mH
Input filter capacitance	C_f	20 μF
Input filter damping resistor	R_d	50 Ω
Output filter inductance	L_o	2 mH
Output filter capacitance	C_o	20 μF

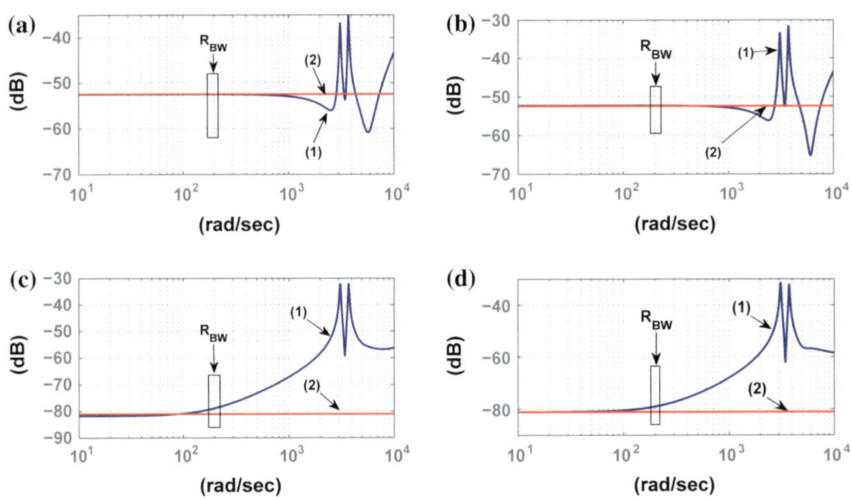

Fig. 4.2 Plot (1): $\mathbf{N_{ff}}(s)$, Plot (2): $\mathbf{A_N}(s)$. **a**, **b** Diagonal elements. **c**, **d** Off-diagonal elements

The parameter values listed in Table 4.2 are used for the present analysis. The input filter parameters do not comply with all the design guidelines provided in Chap. 3. Here the objective behind selecting parameters of Z_{so} was to ensure that P_{cr}, described by (2.106), falls well within the rated power of the experimental prototype MC.

Figure 4.2, shows the frequency response of the elements of $\mathbf{N_{ff}}(s)$ and the zero order approximation $\mathbf{A_N}(s)$. R_{BW} represents the region around the bandwidth frequency. It is evident, the DC approximations almost overlaps with the corresponding elements of $\mathbf{N_{ff}}(s)$ up to ω_{BW}. This is an expected result given the fact that filter design guidelines demands an almost unity gain characteristics from both input and output filters till the fundamental frequency. Hence, upto ω_{BW}, which is even lower than the fundamental frequency, a reasonably close agreement between corresponding elements of $\mathbf{N_{ff}}(s)$ and $\mathbf{A_N}(s)$ is expected. Therefore, the zero order approximation $\mathbf{A_N}(s)$ is finalised as the low frequency approximation. For higher bandwidth requirements, DC approximations may not be sufficient and higher order approximation would become necessary.

In (4.16), although Q_o is a load based parameter, there are methods to control it, as would be discussed in the next chapter. This makes both ω_{co} and Q_o design choices. Therefore compliance to filter design criteria implies that the condition described as

$$\omega_T / (\omega_{co} Q_o) << 1, \tag{4.17}$$

always holds true. Consequently, the diagonal elements of $\mathbf{A_N}(s)$ i.e. $1/a_V$ dominates its dynamics within ω_{BW}.

4.2.1.1 Effects of Variation in Parameters and Operating Points

Any variation in filter parameters is unlikely to affect decoupling process since the corner frequencies are much higher than ω_{BW}. Recalling the expression of a_V from (2.71),

$$a_V = \sqrt{\frac{3}{2}} \hat{V}_{cf} \cos \varphi_i, \tag{4.18}$$

φ_i being an input is known and $\hat{V}_{cf}$ is primarily decided by the source voltage v_s. Since a_V is known, the decoupling matrix $\mathbf{A_N}(s)$ can be modified online corresponding to variations in $\hat{V}_{cf}$. However, this renders the compensator and thereby the control input a function of the state v_{cf} and therefore the dynamic model becomes nonlinear. The alternative route is to use the nominal v_{cf} in $\mathbf{A_N}(s)$. Any variation in $\hat{V}_{cf}$ would lead to a deviation of $\mathbf{N_{ff}^{-1}}(s)\big|_{s=0} \mathbf{A_N}(s)$ from an identity matrix. However, the resulting matrix would still remain a diagonally dominant matrix. As an example, let there be a variation in v_{cf} and filter parameters from their nominal and designed values. Let the modified values of ω_{co}, Q_o and a_V be denoted as $\omega_{co}^{'}$, $Q_o^{'}$ and $a_V^{'}$ respectively. Consequently, the DC gain of $\mathbf{N_{ff}^{-1}}(s)$ is

$$\mathbf{N_{ff}^{-1}}(s)\big|_{s=0} \approx a_V^{'} \begin{bmatrix} 1 & -\omega_T / \left(\omega_{co}^{'} Q_o^{'}\right) \\ \omega_T / \left(\omega_{co}^{'} Q_o^{'}\right) & 1 \end{bmatrix}. \tag{4.19}$$

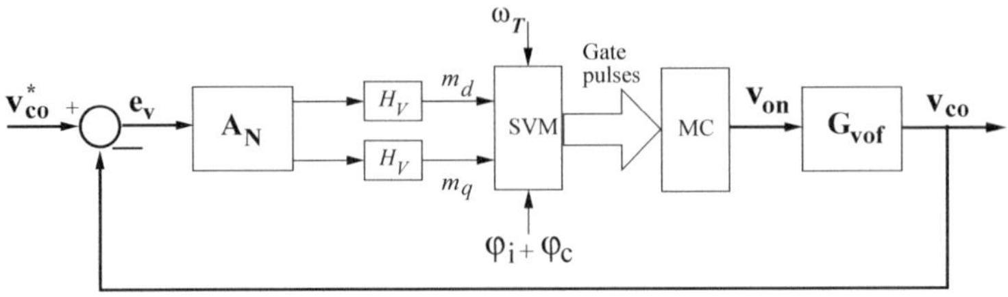

Fig. 4.3 Control scheme

Then

$$\mathbf{N}_{\mathbf{ff}}^{-1}(s)\big|_{s=0}\mathbf{A}_{\mathbf{N}}(s) \approx \frac{a'_V}{a_V}\begin{bmatrix} 1+\dfrac{\omega_T^2}{\omega_{co}Q_o\omega'_{co}Q'_o} & \dfrac{\omega_T}{\omega_{co}Q_o} - \dfrac{\omega_T}{\omega'_{co}Q'_o} \\ \dfrac{\omega_T}{\omega'_{co}Q'_o} - \dfrac{\omega_T}{\omega_{co}Q_o} & 1+\dfrac{\omega_T^2}{\omega_{co}Q_o\omega'_{co}Q'_o} \end{bmatrix}, \tag{4.20}$$

where for any realistic variation in filter parameters the relation

$$\omega_T/\left(\omega'_{co}Q'_o\right) << 1, \tag{4.21}$$

would still hold true. Thus, (4.20) is effectively modified to

$$\mathbf{N}_{\mathbf{ff}}^{-1}(s)\big|_{s=0}\mathbf{A}_{\mathbf{N}}(s) \approx \frac{a'_V}{a_V}\begin{bmatrix} 1 & 0 \\ 0 & 1 \end{bmatrix}. \tag{4.22}$$

Hence, variations in a_V essentially introduces an additional DC gain a'_V/a_V, in the compensator output. The controller can now be designed so that the settling time requirements are met under the expected worst case variation in v_s.

Figure 4.3 shows the employed control scheme. In this regulator problem, a PI algorithm has been chosen for the controller $H_v(s)$. For the controller design, the operating point where input power (P_{in}) is 5 kW has been considered. Plot (a) in Fig. 4.4 shows the frequency response of the plant and plot (b) indicates the loop gain $M_a(s)H_v(s)$. The loop gain has a gain crossover at 200 rad/s. The phase margin of 90° and roll off rate of −20 dB/decade which spans almost a decade on both sides of ω_{BW}, suggest, a first order response characteristic with a settling time of 20 ms. Table 4.3 lists the controller and compensator parameters.

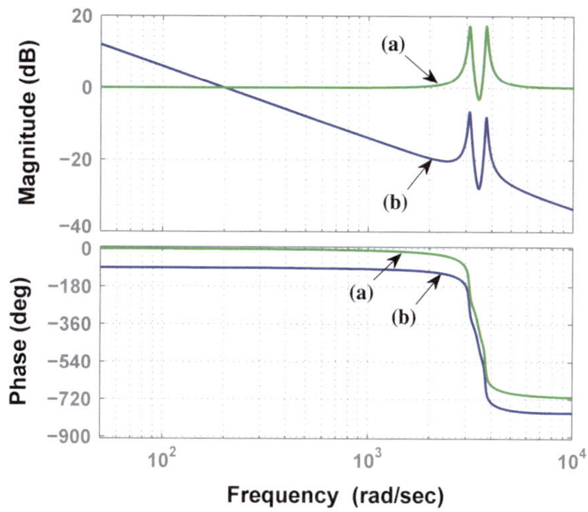

Fig. 4.4 Plant and loop gain. **a** $M_a(s)$ **b** $M_a(s)H_v(s)$

Table 4.3 Compensator and controller parameters

Parameters	Value
H_v	$K_P = 0.0056$, $K_I = 200$
Elements of $\mathbf{A_N}(s)$	0.0023, 0.0001

4.3 Experimental Results and Discussion

The experimental setup described in Chap. 2 has also been used here. Internal resistances of the filter elements were measured at the input corner frequency. These values were then used together with filter parameters tabulated in Table 4.2 for recalculating $|Z_{so11}(j\omega_M)|$ in (2.106). So, for unity IDF, P_{cr} was evaluated as

$$P_{cr} = \frac{3}{2}\frac{\hat{V}_{cf}^2}{|Z_{so11}(j\omega_M)|} = 3.25\,\text{kW}. \tag{4.23}$$

Figure 4.5a shows the dynamic response of one of the phases of output voltage v_{co} following a step reference of 170 V. The nature of the transient response can be better appreciated from the waveform of the modulation index m, which, therefore is provided along with v_{co}. It is observed that m rises from zero and settles exactly at the specified 20 ms demonstrating a first order response characteristic. Similar response is observed when the voltage reference is reset to zero, shown in Fig. 4.5b.

These responses validate the analysis where a first order response with 20 ms response time was predicted. Therefore, this also validates the dynamic model of the overall system from where the reduced order plant was derived.

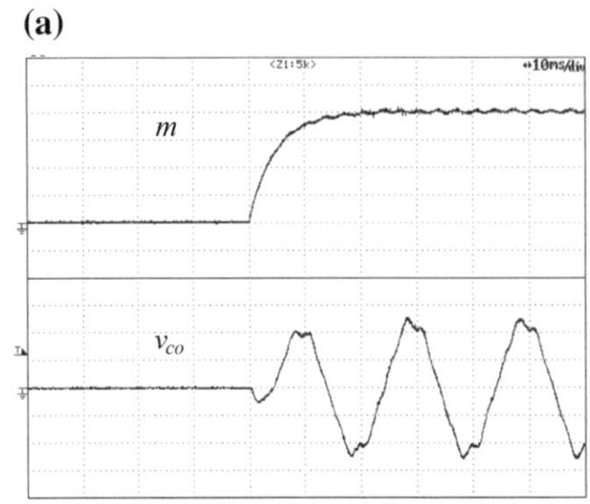

Fig. 4.5 Dynamic response of v_{co}. **a** Step reference of 170 V **b** reference reset to zero. m, scale: 0.2 /div. v_{co}, scale: 100 V/div. Time: 10 ms/div

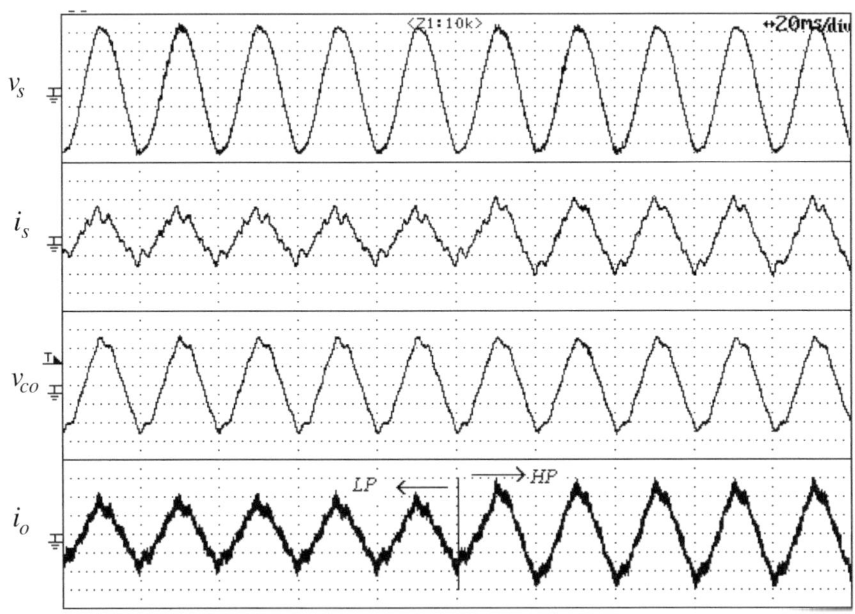

Fig. 4.6 Response to a step change in load. Output current i_o stepped up from 5.3 to 8 A. Source voltage v_s (100 V/div), i_s (5 A/div), v_{co} (100 V/div), i_o (5 A/div). Time: 20 ms/div

Table 4.4 Power delivered to load—measured

i_o (A)	P_L (kW)
5.3	2.62
8	3.95

Table 4.5 Significant harmonic components of experimental waveforms

	v_s (%)	i_s (LP) (%)	i_s (HP) (%)	i_o (LP) (%)	i_o (HP) (%)
3rd	1.3	3.5	3	2.6	2.25
5th	1.75	15	9.71	5	3.4
7th	1.15	12.67	6.7	9.7	6.3
9th	0.55	6	3.63	1.77	1.11
11th	0.55	4.74	2.4	1	0.6
13th	0.55	3.71	2.66	0.4	0.2
THD	3	22	13	11.5	7.6

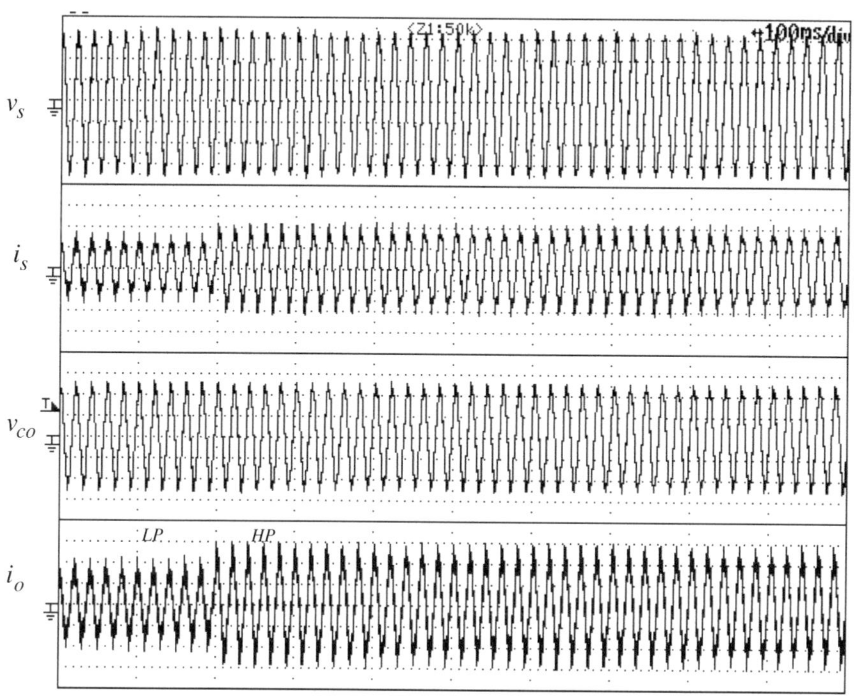

Fig. 4.7 Response to the step change in load shown over a larger period of time. v_s (100 V/div), i_s (5 A/div), v_{co} (100 V/div), i_o (5 A/div). Time: 100 ms/div

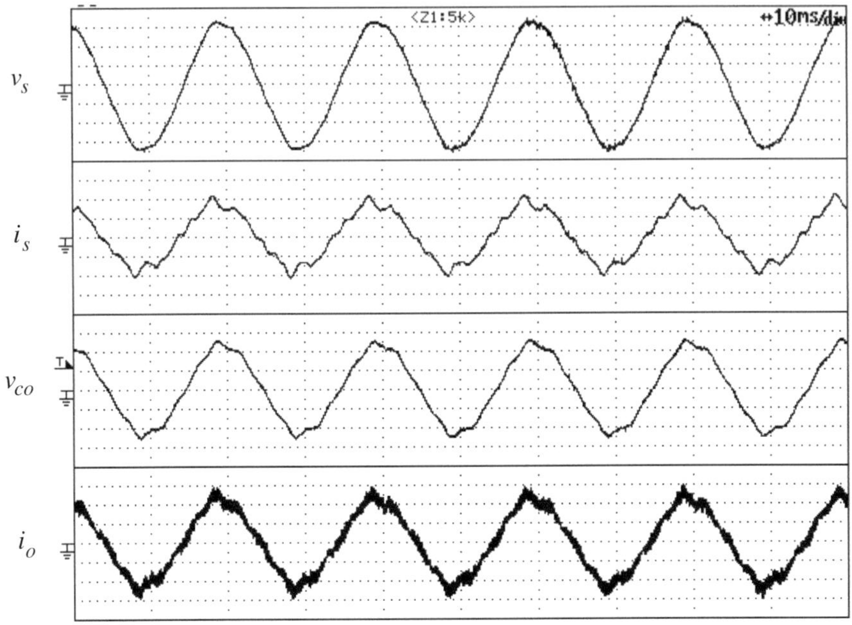

Fig. 4.8 Steady state plots while operating in the HP region. v_s (100 V/div), i_s (5 A/div), v_{co} (100 V/div), i_o (5 A/div). Time: 10 ms/div

A step change in load is then applied which raises the output current, i_o from 5.3 to 8 A. This is shown in Fig. 4.6 where it is evident that load addition has no noticeable effect on the regulated output voltage, v_{co}. This establishes the disturbance rejection capability of the controller. The power delivered to load (P_L), corresponding to the two values of i_o were measured and are tabulated in Table 4.4. The two operating regions are designated as LP and HP. It should also be noted, that the throughput power (P_{in}) in the HP region was higher than the listed P_L, given the finite converter and output filter losses. Therefore, from (4.23) and Table 4.4, P_{in} was well beyond the critical power P_{cr}, at this region of operation.

Harmonic analysis of MC, being outside the scope of this book chapter, is not addressed in detail here. However, qualitatively, harmonic components of v_s affects i_s in two ways. First, the linear input filter introduces harmonics in i_s, specifically due to the capacitance C_f which draws harmonic current. The consequent distortion of the envelope of v_{on} waveform produces low order harmonics in i_o, which are reflected in the load component (i_{in}) of i_s. Table 4.5 lists the dominant harmonic components for the two operating regions shown in Fig. 4.6.

It is important to observe that the harmonic distortions in i_s, *reduces* considerably with the increase in load. Since the capacitive component of i_s is decided by input voltage distortions, it remain the same. Therefore, it is concluded that there is no harmonic amplification in the MC input current (i_{in}) with increasing throughput power. The overall reduction in the harmonic components with greater power transfer rules out any possibility of instability arising out of these low order harmonics.

Figure 4.7 shows the same experimental variables depicted in Fig. 4.6, after the load step, over a larger interval. It is clearly evident that neither input nor output variables show any sign of divergence while operating in the HP region i.e. where the plant is a non-minimum phase one. Figure 4.8 shows the waveforms at steady state along with supply voltage.

4.4 Concluding Remarks

The process of deriving a scalar reduced order plant for controller design has been detailed. This allows application of single variable based techniques for controller design. The inclusion of non-minimum phase zeros in the reduced plant also ensures that internal stability is not violated in the decoupling process. The dynamic response in the experiments shows a very close agreement with the predicted response characteristic. This conclusively validates the control scheme based on model reduction as well as the overall dynamic model from where the reduced plant model is obtained. The converter is made to supply power higher than P_{cr} and is found to display stable operation.

Chapter 5
Voltage Sag Mitigation

Performance of Matrix converter as a voltage sag compensator for linear loads has been described in this chapter. In power quality (PQ) parlance, the term voltage sag is used to designate a dip in the supply voltage magnitude over a small time duration which can extend over few cycles [48, 49]. The importance attached to the quality of power has been increasing rapidly from the later half of the last century. Many reasons have been cited for its increasing significance. With the modernization of power systems, electrical power delivery evolved into a stage where availability of power was increasingly being perceived as a facility which all consumers are entitled to. The last century also witnessed a huge penetration of electronic and power electronic (PE) equipments into power system. These loads were very sensitive to voltage variations (for e.g. microprocessors). Many of these like loads were also responsible for deterioration of PQ. For examples, rectifiers having diode bridge rectifier in the front end where the input current drawn from the grid has significant low frequency harmonics. These called for stricter PQ standards which in turn led to emergence of various techniques for mitigation of PQ problems. These techniques involved use of PE equipments and soon there was a wide scope of PE in power system applications.

One of the key PQ issues is voltage sag. Ofcourse a power interruption has much more severe impact than the transitory voltage sag. However, the overall improvement in power systems has rendered long term interruptions rare in developed industrialized nations. Voltage sags are a much more frequent phenomenon and overall they have more damaging economical impact for the industrial customers. Consequently sag mitigation techniques have high importance at present.

Since the MC has a 3 wire output, it can only be used for compensation of positive and negative sequence voltages. So a 3-phase 3-wire load has been considered here. The presented analysis is focused on the distribution end of power system where most of the loads are inductive in nature. This allows use of the same linearized model described in Chap. 2. The control approach described in Chap. 4 is followed here with necessary modifications for the present application. The topology, voltage injection method chosen and controller design are discussed. Finally, experimental results are presented.

A. Dasgupta and P. Sensarma, *Design and Control of Matrix Converters*,
Energy Systems in Electrical Engineering, DOI 10.1007/978-981-10-3831-0_5

5.1 Topology and Voltage Injection Method

Figure 5.1 shows the topology where MC is used as a Dynamic Voltage Restorer (DVR). Topologically, the structure is very similar to an Unified Power Quality Controller (UPQC) and Unified Power Flow Controller (UPFC). In both applications, active power required during series voltage injection is extracted from the grid through the shunt branch. Traditionally these have been realized using back to back VSIs coupled by the DC link capacitor. Limited life time and the reduced power density due to the bulky electrolytic capacitors have inspired use MCs as UPFCs [40]. The performance objectives of both UPFC and UPQC are defined primarily by regulation of variables in steady state. On the other hand since voltage sags are themselves a transient phenomenon, the transient performance is the main concern for a sag mitigating device.

Input side of MC is connected to the point of common coupling (PCC) and the filtered output voltages (v_{inj}) are injected through transformers. Figure 5.2 shows the single line diagram of the system. Here the PCC voltage is denoted as v_p and its nominal value as v_p^N.

Figure 5.3 shows the three type of voltage injection strategies recommended in literature for sag mitigation using VSIs [50]. v_{ps} and v_{sg} are the PCC voltages before and during sag respectively. In the first method, the injected voltage (v_1) is in quadrature with i_L and hence only reactive power is injected into the system by the DVR. Since active power exchange is not necessary, the energy storage requirement is minimum with this scheme. As energy storage is not a concern with MC, this approach

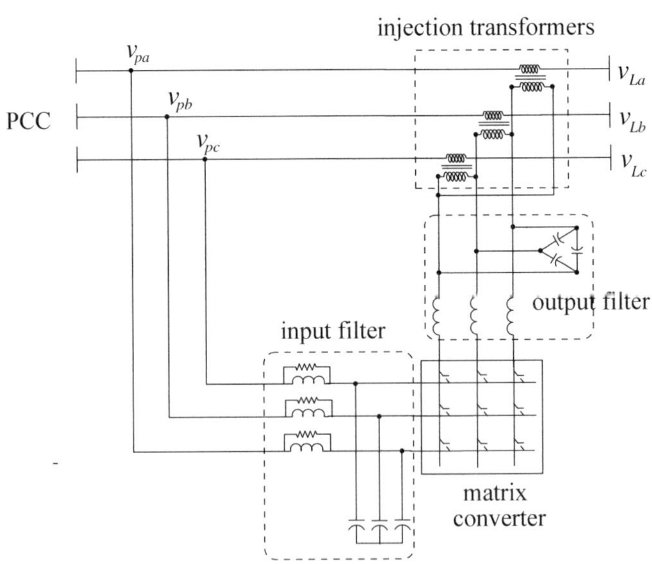

Fig. 5.1 3-phase MC connected as a DVR

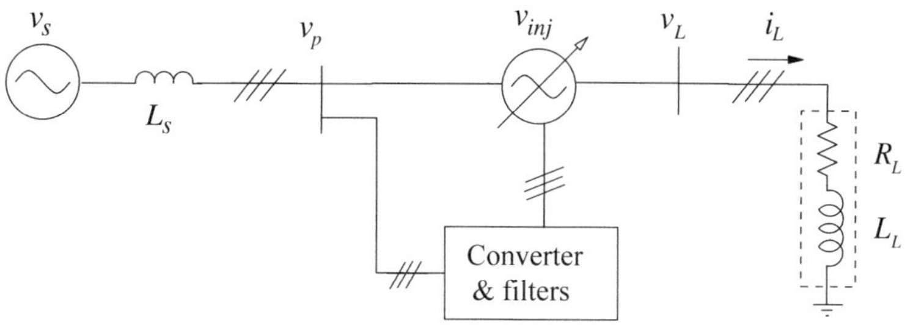

Fig. 5.2 Single line representation of the system

Fig. 5.3 Commonly used compensation strategies

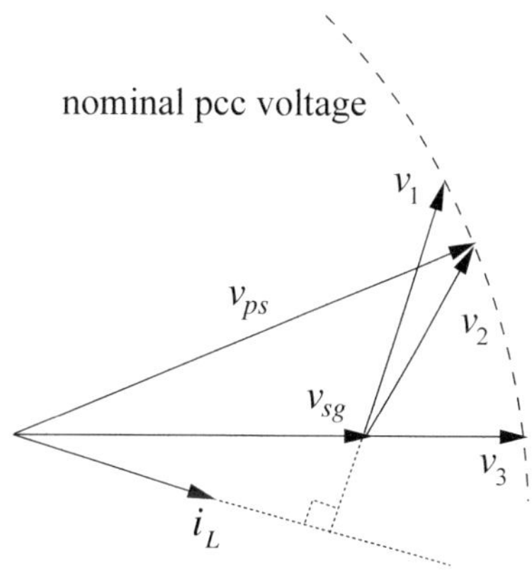

has not been considered in this work. The other two compensation schemes take into consideration both active and reactive power exchange. Objective of voltage injection (v_2) in the second method is to maintain both pre-sag magnitude and phase of the load voltage during sag. This results in minimum disturbance to the load. Realization of this method needs extraction of the phase information of v_{ps}, just before the onset of sag. During sag, the pre-sag phase angle is used for finding the reference command of the injected voltage. The Phase Lock Loop (PLL) which is synchronized to v_{pa} detects the phase of v_{sg} and therefore cannot be used as the reference phase of the injected voltage v_2 during this period. Thereafter, if the post-sag phase of v_p is found to be same as the pre-sag value, then only the PLL output can be used. This is an ideal compensation scheme for some loads which are sensitive to phase jumps in the applied voltage. However, the hardware realization of the scheme involves storage of the pre-sag phase information and therefore requires additional computation and memory resources. In the third method, the injected voltage (v_3) is in phase with v_{sg}. To reduce the effect of phase jump, a slow PLL is used, where, at the starting of sag the load voltage is maintained at v_{ps}. Subsequently as the PLL output synchronizes to v_{sg}, the injected voltage gradually comes in phase with v_{sg}. Hence the phase of the load voltage undergoes a smoother transition. This method allows hardware realization without any additional storage requirement. The restored load voltage here is

$$|v_L| = |v_{sg}| + |v_3|. \tag{5.1}$$

Hence for a specified nominal voltage, the injected voltage magnitude $|v_3|$ is lower than those required for the other two schemes. Since voltage gain available with MC is lower than VSI, the third method is followed in this work to achieve the maximum voltage sag compensation capability.

A PLL reported in [51] extracts the phase of the fundamental positive sequence component in the presence of unbalance or low harmonic distortions. Moreover, the phase locking time is independent of the magnitude of the utility voltage and distortions present in it. This has been used here with a settling time of 16 ms.

The injection voltage reference is initially generated as

$$v_{inj}^{Ref} = v_p - v_p^N. \tag{5.2}$$

In the presence of noise, low frequency harmonics and unbalance associated with any industrial site, it is important to correctly identify an actual voltage sag. Error in sag detection would lead to undesirable injection of series voltage and therefore compounding the problem of load voltage sag or swell instead of rectifying it. For instance, if noise picked up by voltage sensor is falsely interpreted to be a voltage sag, the subsequent voltage injection may result in an overvoltage at the load end. So a hysteresis band has been employed in this work and control is activated when

$$|v_{inj}^{Ref}| \geq 0.12|v_p^N|, \tag{5.3}$$

and subsequently deactivated if

$$|v_{inj}^{Ref}| \leq 0.02|v_p^N|. \tag{5.4}$$

The next objective is to synthesize and realize v_{inj} accurately through MC.

5.2 Plant Model

With predominantly inductive loads in the distribution side, load is modeled as a current stiff element. This leads to the per-phase system shown in Fig. 5.4 where PCC is denoted by **P**.

In this application, the converter is expected to be functional for all kind of load. Hence damping of output filter has to be introduced through the control scheme. This can be achieved in two ways. One such way is to implement a multi loop control strategy where control of i_o is achieved by an inner (faster) loop and v_{co} is controlled by an outer (slower) loop. For regulated voltage supply application in Chap. 4, the given settling time specification for v_{co} of 20 ms translates to a gain crossover of 200 rad/s if the loop gain depicts a first order characteristic. A multi loop control, like the one shown in Fig. 5.5, means that the phase contribution of the inner closed

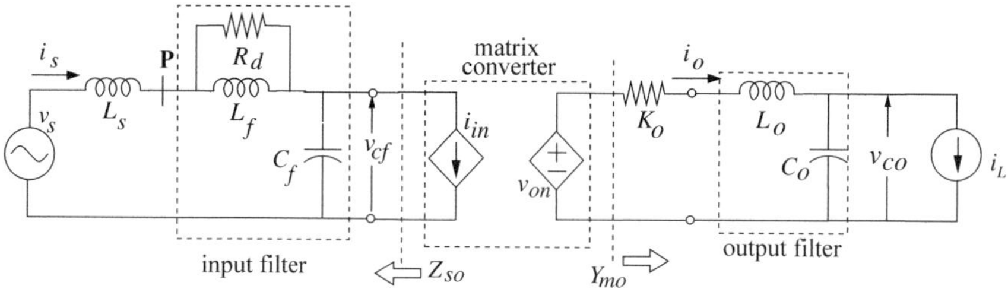

Fig. 5.4 Per phase equivalent circuit

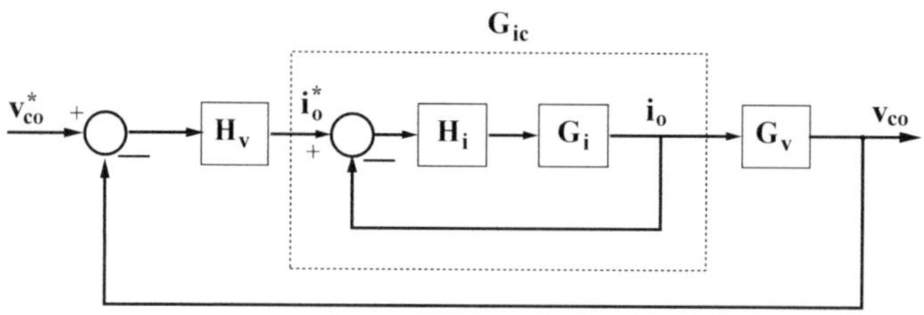

Fig. 5.5 Conventional multiloop control

loop $\mathbf{G_{ic}}$ should be minimum at 200 rad/s. Therefore, the inner loop bandwidth must be much greater than the outer loop bandwidth. For example, if the inner closed loop has a first order dynamics then its bandwidth should be atleast an order of magnitude higher than the outer loop bandwidth. $\mathbf{G_i}$ will be a function of both input and output filter parameters. Consequently, the high gain crossover frequency for the inner loop can now be very close to the corner frequencies of both input and output filters. Controller design becomes very difficult in this situation. The alternative approach is to introduce a virtual damping element where damping is introduced in the system by emulating a resistor through control. This has been followed here.

In Fig. 5.4, K_o is the virtual resistor realized through control as

$$\begin{bmatrix} m_d \\ m_q \end{bmatrix} = H_V \mathbf{A_N} \begin{bmatrix} v_{cod}^* - v_{cod} \\ v_{coq}^* - v_{coq} \end{bmatrix} - \frac{K_o}{v_{pd}^N} \begin{bmatrix} i_{od} \\ i_{oq} \end{bmatrix}. \tag{5.5}$$

Since the aim is to damp output filter resonance, this element is realized along with a first order high pass filter (HPF) as shown in Fig. 5.6. Corner frequency of the HPF is chosen atleast one decade below close loop bandwidth (ω_{BW}) of the system. Therefore the damping transfer function behaves as a constant gain K_o around ω_{BW}. Hence the overall damping transfer function has been modelled by its high frequency gain K_o only.

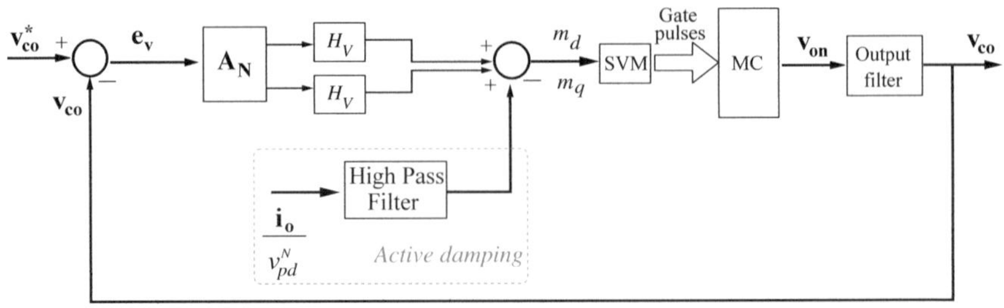

Fig. 5.6 Control scheme with active damping

Here, the reduced scalar transfer function $M_a(s)$ is chosen as

$$M_a(s) = \frac{1}{a_v^2} \frac{det(\mathbf{G_a}(s))}{C_d(\mathbf{G_{vof}}(s))}, \tag{5.6}$$

where the common denominator of $\mathbf{G_{vof}}(s)$ is included to decide on the value of damping component K_o. Using (2.70),

$$\frac{1}{a_v^2} det(\mathbf{G_a}(s)) = 1 - \frac{P_{in}}{\frac{3}{2}\hat{V}_{cf}^2 \cos^2 \varphi_i} Z_{soll}. \tag{5.7}$$

From Fig. 5.4,

$$G_{vof}(s) = \frac{1}{s^2 L_o C_o + s C_o K_o + 1}. \tag{5.8}$$

Hence using (2.83), (2.85) and (5.8),

$$C_d\left(\mathbf{G_{vof}}(s)\right) = \left\{L_o C_o\left(s^2 - \omega_T^2\right) + s C_o K_o + 1\right\}^2 + \omega_T^2 C_o^2 \left\{2 s L_o + K_o\right\}^2. \tag{5.9}$$

Controller design for the plant $M_a(s)$ is described in the next section.

5.3 Control Scheme for DVR

The lowest sag tolerance time for equipments sensitive to voltage dips has been reported to be close to 20 ms for severe voltage sags for more than 50% of nominal voltages [52]. For sags of about 30% of nominal voltage, the lowest tolerance time is about 40 ms. Keeping in mind the tolerance durations, the loop gain of the positive sequence controller block has been designed to have a gain crossover of 200 rad/s and a phase margin of 90°. Separate feedback (FB) control blocks are designed for compensation of positive and negative sequence components.

5.3.1 FB Controller

The precompensator $\mathbf{N}_{\mathbf{G}}^{-1}(s)$ is obtained as

$$\mathbf{N}_{\mathbf{G}}^{-1}(s) = \mathbf{N}_{\mathbf{a}}^{-1}(s)\mathbf{G}_{\mathbf{I}}(s)\left(\mathbf{N}\left(\mathbf{G}_{\mathbf{vof}}\right)(s)\right)^{-1}. \tag{5.10}$$

The DC gain approximation $\mathbf{A}_{\mathbf{N}}$ has been used to realize $\mathbf{N}_{\mathbf{G}}^{-1}(s)$. Denoting the fundamental and switching frequency as f_b and f_s respectively, the nominal parameters given in Table 5.1 are considered for controller design. K_o has been selected to attain a quality factor (Q_o) of 2.3, for the filter parameters given in Table 5.2. This value of Q_o, apart from introducing necessary damping also ensures sufficient switching ripple attenuation. C_o is the effective capacitance value that appears in per phase modelling.

A PI controller $(H_P(s))$ has been used for positive sequence components with a target settling time of 20 ms. Figure 5.7 shows the plant and the loop gain.

The negative sequence components appear as 2nd harmonic components in synchronous dq frame. This is clarified with the following set of 3-phase variables,

$$X_a = X_m\cos(\omega_T - \theta),\ X_b = X_m\cos(\omega_T + 120^\circ - \theta)\ \&\ X_c = X_m\cos(\omega_T - 120^\circ - \theta). \tag{5.11}$$

Using the transfer matrix $\mathbf{T}$ defined in (2.32) as

$$\mathbf{T} = \sqrt{\frac{2}{3}}\begin{bmatrix} \cos(\omega_T t) & \cos(\omega_T t - 120^\circ) & \cos(\omega_T t + 120^\circ) \\ \sin(\omega_T t) & \sin(\omega_T t - 120^\circ) & \sin(\omega_T t + 120^\circ) \\ 1/\sqrt{2} & 1/\sqrt{2} & 1/\sqrt{2} \end{bmatrix}, \tag{5.12}$$

the variables in (5.11) are converted to

$$X_d = \sqrt{\frac{3}{2}}X_m\cos(2\omega_T - \theta)\ \&\ X_q = \sqrt{\frac{3}{2}}X_m\sin(2\omega_T - \theta). \tag{5.13}$$

Table 5.1 Nominal parameters

f_b	V_s	f_s	L_s	R_L
50 Hz	230 V	10 kHz	3 mH	120 Ω

Table 5.2 Input and output filter parameters

C_f	L_f	R_d	C_o	L_o
15 μF	1.26 mH	25 Ω	15 μF	2 mH

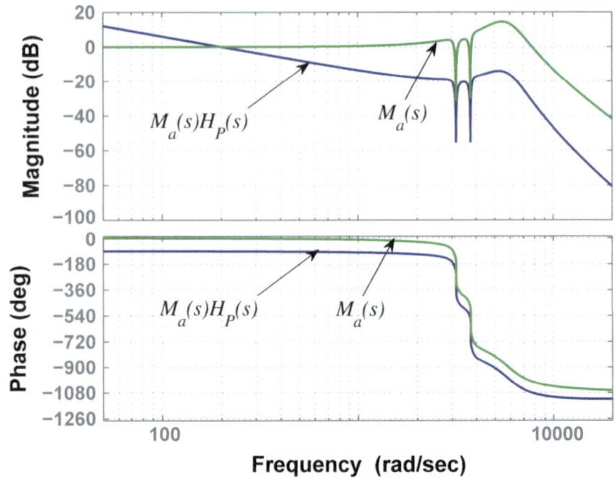

Fig. 5.7 Bode plot of $M_a(s)$ and $M_a(s)H_P(s)$

These 2nd harmonic components can be compensated by redesigning $H_P(s)$ shown in Fig. 5.7. However this would require a very high gain crossover frequency, which makes the attenuation of the gain at the filter crossover frequencies difficult. Without modifying $H_P(s)$, the task of compensating this single frequency component can be accomplished by using an additional controller of the form [53],

$$H_R(s) = K_R \frac{s/\omega_2}{(s/\omega_2)^2 + (s/Q_R\omega_2) + 1}. \tag{5.14}$$

The idea behind using $H_R(s)$ is to introduce a gain over a narrow frequency band surrounding its corner frequency. The corner frequency is selected at ω_2, which is the 2nd harmonic frequency. Bode plot of $H_R(s)$ is shown in Fig. 5.8.

Q_R and K_R are the design parameters in (5.14). Higher Q_R increases the resonant peak and therefore lowers steady state error. However this leads to sharper phase

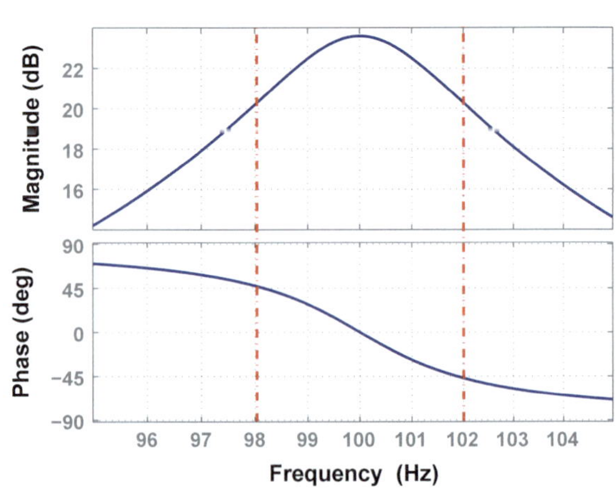

Fig. 5.8 Bode plot of $H_R(s)$

Table 5.3 Parameters of $H_R(s)$

ω_2	Q_R	K_R
628 rad/s	27	0.56

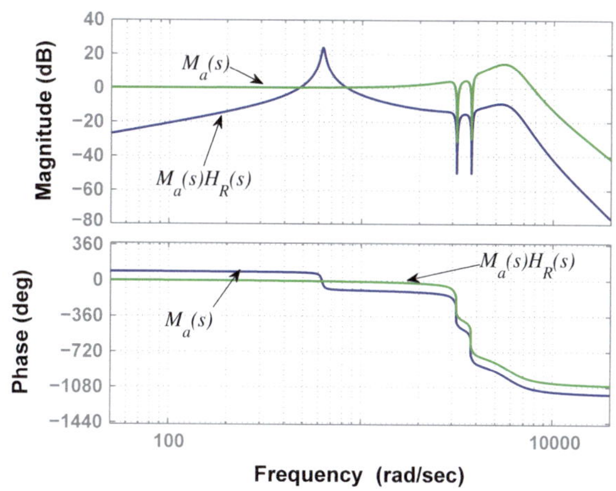

Fig. 5.9 Bode plot of $M_a(s)$ and $M_a(s)H_R(s)$

transition around ω_2 making the circuit more selective to a particular value of ω_2. At the experimental site, f_b is found to vary within ± 1 Hz of its nominal value given in Table 5.1. $H_R(s)$ has been realized corresponding to the nominal value of f_b. To account for the possible variations in frequency, $K_R Q_R$ is selected to ensure a steady state tracking error in magnitude and phase less than 10% and 45° respectively at 98 and 102 Hz. Figure 5.8 depicts this and the selected values of K_R and Q_R are shown in Table 5.3.

The frequency response of the plant and the loop gain $M_a(s)H_R(s)$ is shown in Fig. 5.9. $H_P(s)$ and $H_R(s)$ together results in the FB Controller

$$H_v(s) = H_P(s) + H_R(s), \tag{5.15}$$

Simulation results with the FB controller is shown in Fig. 5.10. Sag was introduced in phase-a (plot in red) of the PCC voltages at the instant T_B and was removed at the instant T_E. Both sag mitigation and subsequent disengagement of the voltage injection at the time of sag recovery is achieved within the target 20 ms period. However at the instant of sag recovery, T_E, the PCC voltage of phase-a was at its positive peak. This PCC voltage together with the injected voltage resulted in over-voltage at the load side. While trying to perform the experimental tests, an over-voltage fault at the load occurred whenever sag was removed from the system. Maintaining uninterrupted operation in this situation demands a very high ω_{BW} for the FB controller. This can be particularly difficult to achieve if throughput power exceeds P_{cr} resulting in the emergence of nonminimum phase zeros. Therefore a

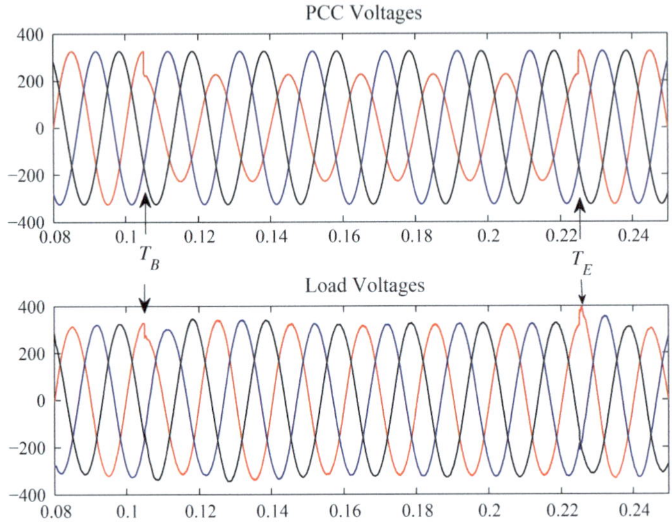

Fig. 5.10 Simulation result with FB controller. Phase a *red*, b *blue*, c *black* (colour online)

Feed Forward (FF) controller is used along with the FB controller to reduce the severity of this over-voltage. This is described in the next section.

5.3.2 FF Controller

A FF controller in the form

$$m_{fd} = \frac{v^*_{cod}}{v^N_{pd}}, \quad m_{fq} = \frac{v^*_{coq}}{v^N_{pd}} \tag{5.16}$$

has been used. As evident, the ratio of reference output to nominal PCC voltage is used to generate the control output of the FF controller. Use of nominal PCC voltage rules out any chance of instability that might emerge from instantaneous compensation as reported in the conventional FF control approaches for MC [8]. The FF controller employed here basically works as an open loop compensator.

Figure 5.11 shows the overall control scheme. Experimental results are presented in the following section.

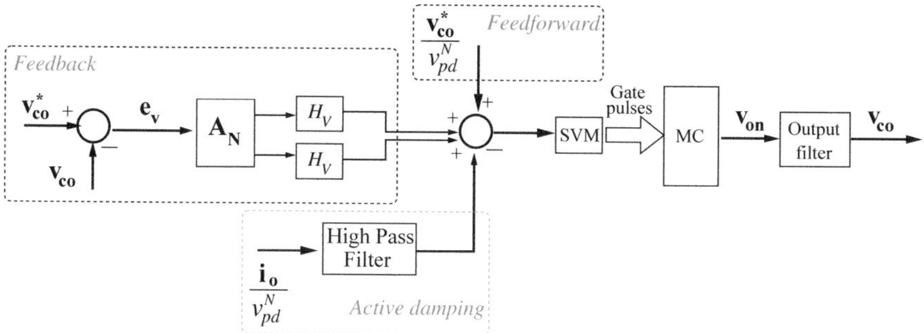

Fig. 5.11 Control scheme

5.4 Experimental Results and Discussion

Since load is considered to be current stiff in analysis, the analytically developed model is closest to the experimental model for a dominantly inductive load. Moreover, a stiffer i_L, which is the disturbance source to the output filter as shown in Fig. 5.4, implies lower possibilities of oscillation. So, if satisfactory dynamic performance is obtained with only resistive load, it guarantees similar, if not better performance with any other linear load. Hence a 3—wire star connected resistive load has been considered in this work.

Figure 5.12 shows the single phase diagram of the circuit used to generate sag in v_p. C1 and C2 are two mechanically interlocked, normally open and closed contactors, respectively. When C1 is energized, a sag is initiated.

Nominal voltage given in Table 5.1 was treated as the load voltage reference. Voltage sag of about 25% was generated for both symmetrical and asymmetrical sag cases.

Figure 5.13 shows the PCC and load voltages for symmetrical sag. The sag is initiated when phase-a (plot in yellow) of PCC voltage is at its peak as seen in Fig. 5.13a and the load voltages at that instant are shown in Fig. 5.13b where they are seen to be restored within 10 ms. Removal of sag takes place at the instant when one of the PCC voltage is approaching its negative peak as observed in Fig. 5.13c. This is a situation, where, the FF controller becomes necessary. The voltage overshoot in the corresponding phase of v_L at this instant is measured to be 120 V, as shown in Fig. 5.13d. Instantaneous compensation using the FF controller is difficult to achieve

Fig. 5.12 Sag generation setup shown for one phase

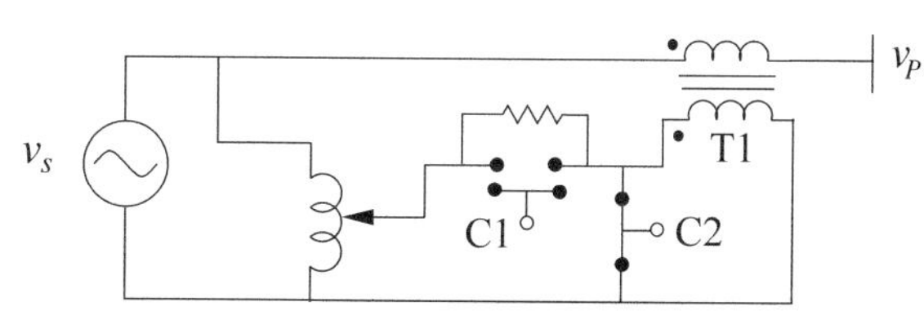

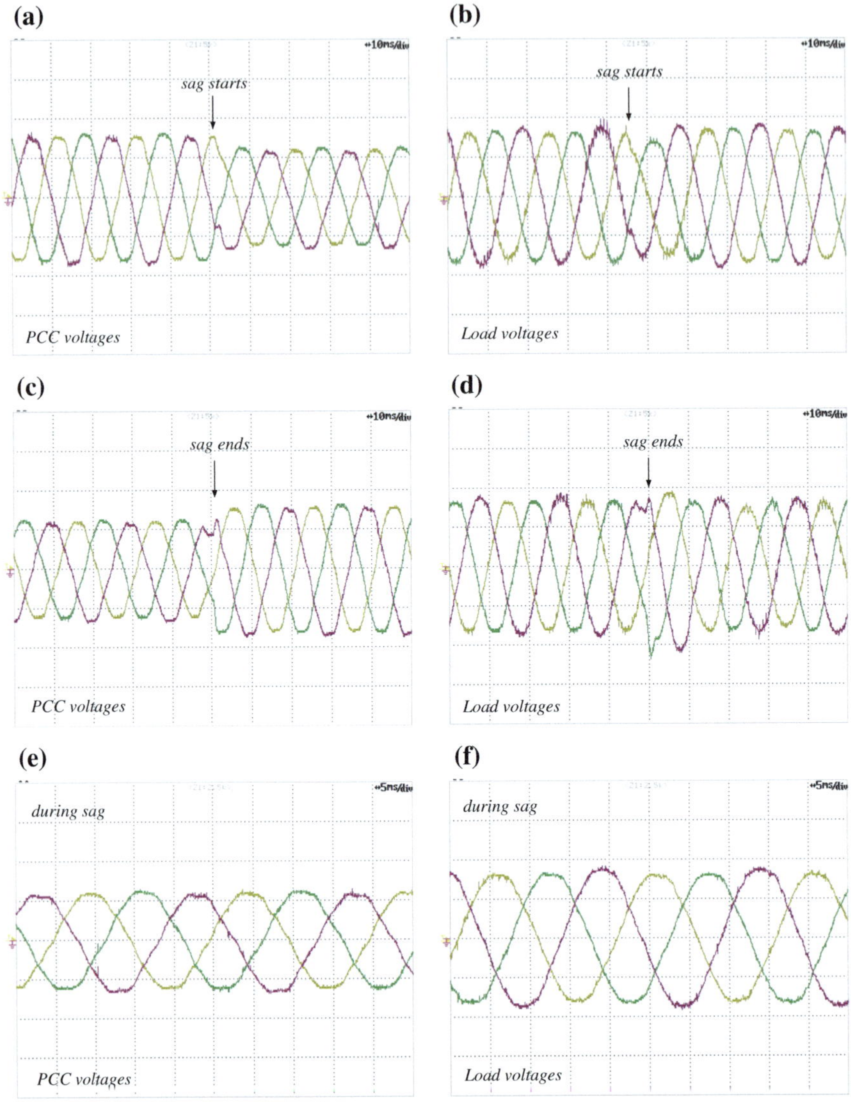

Fig. 5.13 Experimental waveforms for symmetrical voltage sag. Phase a *yellow*, b *green*, c *purple*. Voltage scale: 200 V/div. Time: 10 ms/div for (**a**)–(**d**) and 5 ms/div for (**e**) and (**f**) (colour online)

Table 5.4 Harmonic components during symmetrical sag

	3rd, %	5th, %	7th, %	9th, %	11th, %	13th, %	THD, %
v_{pa}	1.28	1.64	1.71	1.23	1.56	1.17	4.6
v_{La}	1.07	1.62	1.37	0.44	0.43	X	3.44
v_{pb}	0.9	2.51	2.16	0.69	0.65	X	4.66
v_{Lb}	0.42	2.02	1.39	0.4	X	X	3.43
v_{pc}	1.19	2.78	2.69	1.14	0.833	0.75	5.28
v_{Lc}	0.89	1.7	1.15	0.3	0.4	X	3.7

Table 5.5 Harmonic components before initiating symmetrical sag

	3rd, %	5th, %	7th, %	9th, %	11th, %	13th, %	THD, %
v_{pa}	1.2	1.9	1.56	1.07	1.16	0.87	3.8
v_{La}	1.67	1.04	0.94	0.31	0.83	X	4.05
v_{pb}	1.04	2.51	2	0.7	0.67	X	4.04
v_{Lb}	1.24	1.57	1.67	0.4	X	0.4	3.6
v_{pc}	0.9	2.44	2.75	1.02	0.58	0.58	4.45
v_{Lc}	1.9	1.8	1.62	X	0.7	0.5	4.3

in presence of the unavoidable delays introduced by measurement as well as digital control hardware. Still, the over-voltage is mitigated well within 5 ms. A distortion near the positive peak of another phase of PCC voltage can also be observed in Fig. 5.13c. This is due to the finite interval between opening of C1 and closing of C2, during which the current stiff primary side of T1 sees the voltage across variac and resistance across C1 in series. This is however a limitation of the sag generation circuit and not of the voltage controller. The PCC and load voltages during sag are shown in an expanded time scale in Fig. 5.13e, f respectively.

The experimental data were recorded and the harmonic components of the PCC and load voltages during sag is provided in Table 5.4. The corresponding values before sag are listed in Table 5.5. Those harmonic components which were found to be lower than 0.3% are neglected and the corresponding entries have been denoted as 'X'. As evident the 3rd harmonic component in the load voltages is more than their PCC counterparts in normal mode while it is lower during the presence of sag. However the THD of each of the phase voltages remains very much comparable. Hence the effect of the converter on the overall system is minimal during normal operation.

Figure 5.14 shows the PCC and load voltages for asymmetrical sag. Sag was introduced only in phase-a (plot in yellow) of the PCC voltage. The load voltages are seen to be restored within 10 ms at both the start and end of sag shown in Fig. 5.14b, d. The sag removal took place when one phase of PCC was at its positive peak as shown in Fig. 5.14c. The voltage jump in the corresponding load voltage

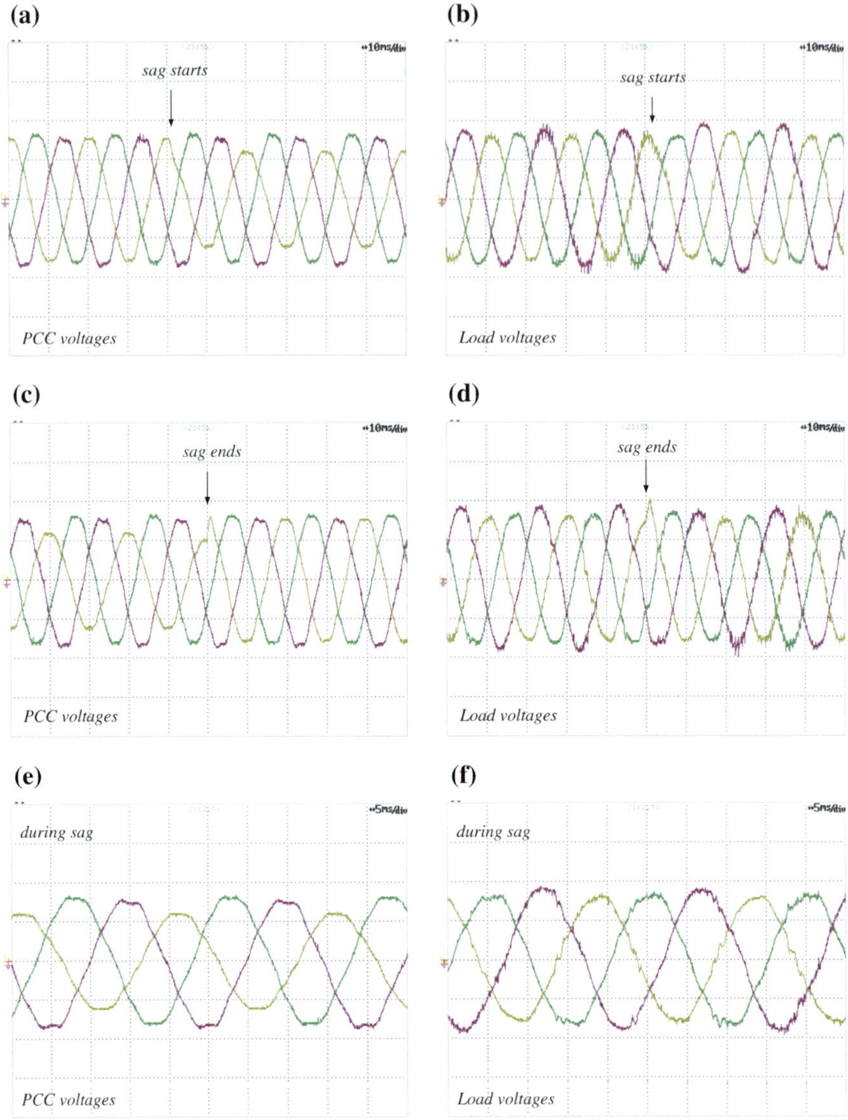

Fig. 5.14 Experimental waveforms for asymmetrical voltage sag. Phase a *yellow*, b *green*, c *purple*. Voltage scale: 200 V/div. Time: 10 ms/div for (**a**)–(**d**) and 5 ms/div for (**e**) and (**f**) (colour online)

shown in Fig. 5.14d was measured to be 90 V. The PCC and load voltages during sag are shown in Fig. 5.14e, f.

The positive and negative sequence components were extracted from the experimental data using sequence analyzer tool in MATLAB corresponding to the prevalent fundamental frequency. Peak value of these components are shown in Table 5.6. It

Table 5.6 Positive and negative sequence components in PCC and load voltage before and during asymmetrical sag

	v_{p+} (V)	v_{p-} (V)	v_{L+} (V)	v_{L-} (V)
Before sag	325	5	334	10
During sag	300	29	335	15

Table 5.7 Harmonic components during asymmetrical sag

	3rd, %	4rth, %	5th, %	6th, %	7th, %	9th, %	11th, %	13th, %	THD, %
v_{pa}	1.65	X	1.4	0.4	1.65	1.32	1	0.9	4.26
v_{La}	3.25	1	0.5	1.14	1.31	X	1	1.22	5.4
v_{pb}	1.4	X	2.3	X	2.13	1	0.66	X	4.3
v_{Lb}	1.5	X	1.8	X	2.06	0.5	X	0.5	4.1
v_{pc}	1.14	X	2.13	0.33	3.31	1	0.76	0.8	5.1
v_{Lc}	4.6	X	1.3	0.47	0.5	0.5	0.8	1	6.2

Table 5.8 Harmonic components before initiating asymmetrical sag

	3rd, %	5th, %	6th, %	7th, %	9th, %	11th, %	13th, %	THD, %
v_{pa}	0.8	1.51	0.4	2	1.15	1.34	1	4.11
v_{La}	1.5	0.76	X	1.6	0.53	0.5	0.5	3.9
v_{pb}	0.77	2.26	X	2.1	0.8	0.88	0.33	4.06
v_{Lb}	0.74	0.68	X	1.9	0.65	0.3	0.31	4
v_{pc}	1	2.04	X	2.8	1.4	0.94	0.6	4.8
v_{Lc}	0.8	1.1	X	1.74	X	0.47	0.3	4.77

is observed that, during sag, positive sequence component in PCC drops to 92% of its presag value, while that in load remains almost the same. The negative sequence during sag increases by almost 500% in PCC and 50% in the load voltage. Steady state error for negative sequence is restricted to 21% which is greater than the 10% limit considered in design. It is difficult to analyze the reason behind this discrepancy since the design was based on a balanced 3-phase system.

The harmonic components of the PCC and load voltages during and before sag are provided in Tables 5.7 and 5.8 respectively. During the sag period the 3rd harmonic in load voltage is noticeably high and even harmonics appear in one of the phases. THD of the load voltages are also higher than the PCC voltage during this period. During normal operation the harmonic component of load and source voltages are almost similar.

5.5 Concluding Remarks

Sag mitigation is demonstrated for both symmetrical and asymmetrical load. The dynamic response time found to be lower than 10 ms for both type of sags which is much lower than the reported sag tolerance time of sensitive equipments. During asymmetrical sag, the harmonic content of the load voltages was found to be higher than their presag values. The cause behind this could not be traced in this Book and remains to be investigated in future. However the major concern in voltage sag is restoration of the load voltage within the power supply hold up times of equipments. In this regard, the demonstrated performance meets the requirement of a competent sag mitigating device. It is also observed that the MC has negligible effect on the load voltage in absence of sag. This is an important requirement as DVRs remain idle most of the time during which its effect on the overall system should be minimal.

Chapter 6
Conclusion

In this work a design approach has been formulated for 3-phase direct Matrix Converters for selected synchronous applications with linear load. In closure, it is important to summarize the conclusions which could be reached in this work and those aspects which needs future investigations. In view of these, some thoughts on today's relevance this work is also presented at the end.

6.1 Conclusions and Contributions

The conclusions and contributions are categorized in the following manner.

Modelling: The presented analysis began with the conventional small signal analysis of the multi variable plant model in synchronous dq domain. Thereafter, every component part of the plant was investigated and the condition under which it becomes a nonminimum phase one was clearly defined.

(i) This provides the system designer a direct insight into the interaction between the physical sub-systems.
(ii) The plant obtained from the chosen input-output variables is stable under all operating conditions as opposed to an unstable one obtained with erst-while control approaches employing feed forward compensation techniques.
(iii) It is clarified, with the example of a generic DC-DC converter having the same equivalent power stage as Matrix Converter, that the power stage and input filter are responsible for the nonminimum phase zeros in the linearized model.

Therefore these observations are equally applicable to any other modulation, modelling technique and also to single phase and indirect Matrix Converters which are derived from the direct 9 switch topology.

Filter design: An extensive filter design approach is presented which in addition to meeting general requirements of a ripple filter also addresses the design constraints

A. Dasgupta and P. Sensarma, *Design and Control of Matrix Converters*,
Energy Systems in Electrical Engineering, DOI 10.1007/978-981-10-3831-0_6

imposed by dynamic specifications and commutation requirements. It has been discussed that for high grid inductance, controller and filter design has to be treated simultaneously. If the bandwidth requirements are high then a trade off between dynamic specifications and voltage regulation, losses may become necessary.

Controller design for regulated voltage supply: Reduction of the plant model, while preserving internal stability, which allows application of single variable based control technique has been detailed. Experimental results validates the accuracy of the dynamic model, model order reduction and controller design. Stable operation is demonstrated in the operating region where nonminimum phase zeros of the plant appear.

Sag mitigation: Mitigation of symmetrical and asymmetrical voltage sag is demonstrated. Dynamic response time is lower than power supply hold-up time for majority of sensitive equipments. Order of the controller is same as that used in VSIs for compensation of positive and negative sequence components and is easily realized on digital platform.

6.2 Scope of Future Work

Some observations were made in Chap. 5 regarding the increment of third and also emergence of even harmonic. It is also not clear about why the negative components in the load voltage could not be contained within the limits specified by design. These may be addressed on an individual basis.

Although, asymmetrical sag mitigation has been attempted, the model has been developed assuming balanced 3-phase system. So the low frequency modelling based on the transfer function approach detailed in Chap. 2 may be extended to include the zero sequence components for 4 leg MCs [54].

A complete harmonic analysis of the modulation methods MCs is still not available. A comparative evaluation of MCs with B2B VSI can be complete only after such an analysis is available. This work has demonstrated that the desirable dynamic performance for some selected power system applications can be met with MC. Harmonic analysis is the necessary next step in order to reach any final conclusion on the merit of using MC in power system applications, particularly for nonlinear loads. So this is an area which can be addressed. This would also enable more informed hardware design, for e.g. selecting the upper limit for low frequency gain of output filter in Chap. 3.

6.3 Relevance of This Work at Present

This research started almost a decade back and spanned well over half of it. During this period, there has been a lot of parallel research activities elsewhere and so the present scenario is of course not exactly the same as before. But a majority of those activities still continue to be oriented towards the low power electrical drives/modulation methods/topologies while a few are trying to experiment with other domains. It is beyond the scope of this book to provide a detailed overview of the same.

From the point of industrial interest, the situation is discouraging. Two (FUJI and YASKAWA Electric) companies among the entire global community were known to have introduced MC based drives in their product line up about a decade back i.e. even before this research was initiated. However, the situation has not improved over these years. From an extensive survey of journals, magazines and white papers there seems to be a perceived apprehension regarding the reliability and complexity of MCs in general. A considerable amount of analysis for comparing MCs with other 3-phase AC-AC converters exists from academia. Some of the notable ones among them are: an exhaustive analysis and comparison of performance for low voltage low power electrical drives in [21, 55] and a summary of the issues unsolved/partially solved in [56]. There is no debate on the fact that B2B VSIs are a superior solution in terms of voltage gain and ease of control. However it has emerged, from these studies, that MCs have a higher power density, power to mass ratio and better efficiency at higher switching frequencies. Some have also concluded that MCs are not necessarily less reliable and the choice of the most reliable converter is application dependent. However, a similar analysis of MCs is still missing from the industry which are much better equipped to conduct these investigations at low and medium power levels. This kind of a participation is necessary from the industry to arrive at some conclusions regarding the almost negligible penetration of MCs in industry.

This study has been successful in providing an insight into the plant model which should definitely enable future systems designs with a greater understanding of the system. Experimental verifications have been carried out in a low voltage, low power prototype. It would gain relevance once these are put to rigorous tests at higher power levels. An industrial setup is definitely much better equipped than an academic lab for this purpose. Today, the knowledge gained from this research together with many other from academia forms an enormous reservoir of knowledge concerning MCs. Browsing the recent editions of some reputed journals tells that academic research on MCs is still very much alive. On the other hand the fact that MCs has not still been able to find their way to industry after almost 4 decades of research is surprising. More disconcerting is the fact that the reasons are not clearly known. It is only by industrial participation this obscureness can be addressed and settled—which is the need of today.

Appendix A
Derivations Associated with Low Frequency Dynamic Model

A.1 Expanded form of $\mathbf{G}_{\mathbf{I}}^{-1}(s)$

From (2.67), $\mathbf{G}_{\mathbf{I}}(s)$ has been used to denote

$$\mathbf{G}_{\mathbf{I}} = \mathbf{I}_2 + \mathbf{M}\mathbf{Z}_{\mathbf{so}}\mathbf{M}^T\mathbf{Y}_{\mathbf{mo}}. \tag{A.1}$$

$\mathbf{M}$ can be obtained using (2.35), (2.36), (2.44) and (2.45) as

$$\mathbf{M} = \frac{\sqrt{3}}{2}m\begin{bmatrix}\cos\varphi_o\\ \sin\varphi_o\end{bmatrix}\begin{bmatrix}C_\theta & S_\theta\end{bmatrix} = \begin{bmatrix}M_d\\ M_q\end{bmatrix}\begin{bmatrix}C_\theta & S_\theta\end{bmatrix}, \tag{A.2}$$

where

$$C_\theta = \cos(\varphi_i + \varphi_c) \quad \& \quad S_\theta = \sin(\varphi_i + \varphi_c).$$

Laplace operator s has been dropped in most of the derivations shown in this appendix to avoid cluttering. Nomenclature used for the transfer matrix elements of $\mathbf{G}_{\mathbf{fv}}$ in (2.55) has also be used for the transfer function matrices $\mathbf{Z}_{\mathbf{so}}$, $\mathbf{Y}_{\mathbf{mo}}$ and later for $\mathbf{G}_{\mathbf{vof}}$ in this book. Subsequently, using (A.2) in (A.1),

$$\begin{aligned}
\mathbf{G}_{\mathbf{I}} &= \mathbf{I}_2 + \begin{bmatrix}M_d\\ M_q\end{bmatrix}\begin{bmatrix}C_\theta & S_\theta\end{bmatrix}\begin{bmatrix}Z_{so11} & Z_{so12}\\ -Z_{so12} & Z_{so11}\end{bmatrix}\begin{bmatrix}C_\theta\\ S_\theta\end{bmatrix}\begin{bmatrix}M_d & M_q\end{bmatrix}\mathbf{Y}_{\mathbf{mo}}\\
&= \mathbf{I}_2 + \begin{bmatrix}M_d\\ M_q\end{bmatrix}\begin{bmatrix}C_\theta Z_{so11} - S_\theta Z_{so12} & C_\theta Z_{so12} + S_\theta Z_{so11}\end{bmatrix}\begin{bmatrix}C_\theta\\ S_\theta\end{bmatrix}\begin{bmatrix}M_d & M_q\end{bmatrix}\mathbf{Y}_{\mathbf{mo}}\\
&= \mathbf{I}_2 + Z_{so11}\begin{bmatrix}M_d\\ M_q\end{bmatrix}\begin{bmatrix}M_d & M_q\end{bmatrix}\begin{bmatrix}Y_{mo11} & Y_{mo12}\\ -Y_{mo12} & Y_{mo11}\end{bmatrix}\\
&= \begin{bmatrix}1 + M_d Z_{so11}\left(M_d Y_{mo11} - M_q Y_{mo12}\right) & M_d Z_{so11}\left(M_d Y_{mo12} + M_q Y_{mo11}\right)\\ M_q Z_{so11}\left(M_d Y_{mo11} - M_q Y_{mo12}\right) & 1 + M_q Z_{so11}\left(M_d Y_{mo12} + M_q Y_{mo11}\right)\end{bmatrix}.
\end{aligned} \tag{A.3}$$

A. Dasgupta and P. Sensarma, *Design and Control of Matrix Converters*, Energy Systems in Electrical Engineering, DOI 10.1007/978-981-10-3831-0

So determinant of $\mathbf{G_I}$ is

$$det\left(\mathbf{G_I}\right) = 1 + \left({M^2}_d + {M^2}_q\right) Z_{so11} Y_{mo11} = 1 + \frac{3}{4} m^2 Z_{so11} Y_{mo11}. \qquad (A.4)$$

Therefore

$$\mathbf{G_I}^{-1} = \begin{bmatrix} 1 + M_q Z_{so11}\left(M_d Y_{mo12} + M_q Y_{mo11}\right) & -M_d Z_{so11}\left(M_d Y_{mo12} + M_q Y_{mo11}\right) \\ -M_q Z_{so11}\left(M_d Y_{mo11} - M_q Y_{mo12}\right) & 1 + M_d Z_{so11}\left(M_d Y_{mo11} - M_q Y_{mo12}\right) \end{bmatrix} \times \frac{1}{1+\frac{3}{4} m^2 Z_{so11} Y_{mo11}}. \qquad (A.5)$$

A.2 $\mathbf{G_a}(s)$

By setting $\tilde{\theta}_i(s)$ to zero in (2.44) and using (2.45), $\tilde{\mathbf{m}}$ can be expressed as

$$\tilde{\mathbf{m}} = \begin{bmatrix} \tilde{m}_d \\ \tilde{m}_q \end{bmatrix} \begin{bmatrix} C_\theta & S_\theta \end{bmatrix} = \begin{bmatrix} \tilde{m}_d \\ \tilde{m}_q \end{bmatrix} \begin{bmatrix} \cos(\varphi_i + \varphi_c) & \sin(\varphi_i + \varphi_c) \end{bmatrix}. \qquad (A.6)$$

Using $\tilde{\mathbf{m}}$ from (A.6),

$$\left(\tilde{\mathbf{m}}\mathbf{V_{cf}} - \mathbf{M}\mathbf{Z_{so}}\tilde{\mathbf{m}}^T\mathbf{I_o}\right) = \underbrace{\begin{bmatrix} \tilde{m}_d \\ \tilde{m}_q \end{bmatrix} \begin{bmatrix} C_\theta & S_\theta \end{bmatrix} \begin{bmatrix} V_{cfd} \\ V_{cfq} \end{bmatrix}}_{\mathbf{A}} - \underbrace{\begin{bmatrix} M_d \\ M_q \end{bmatrix} \begin{bmatrix} C_\theta & S_\theta \end{bmatrix} \begin{bmatrix} Z_{so11} & Z_{so12} \\ -Z_{so12} & Z_{so11} \end{bmatrix} \begin{bmatrix} C_\theta \\ S_\theta \end{bmatrix} \begin{bmatrix} \tilde{m}_d & \tilde{m}_q \end{bmatrix} \begin{bmatrix} I_{od} \\ I_{oq} \end{bmatrix}}_{\mathbf{B}}. \qquad (A.7)$$

Using (2.1) and (2.32),

$$\begin{bmatrix} V_{cfd} \\ V_{cfq} \end{bmatrix} = \sqrt{\frac{3}{2}} \hat{V}_{cf} \begin{bmatrix} \cos\varphi_c \\ \sin\varphi_c \end{bmatrix}. \qquad (A.8)$$

Hence the factor $\mathbf{A}$ in (A.7) can be simplified to

$$\mathbf{A} = \begin{bmatrix} \tilde{m}_d \\ \tilde{m}_q \end{bmatrix} \begin{bmatrix} \cos(\varphi_i + \varphi_c) & \sin(\varphi_i + \varphi_c) \end{bmatrix} \sqrt{\frac{3}{2}} \hat{V}_{cf} \begin{bmatrix} \cos\varphi_c \\ \sin\varphi_c \end{bmatrix} = \sqrt{\frac{3}{2}} \hat{V}_{cf} \cos\varphi_i \begin{bmatrix} \tilde{m}_d \\ \tilde{m}_q \end{bmatrix}. \qquad (A.9)$$

$\mathbf{B}$ in (A.7) simplifies to

$$\mathbf{B} = \begin{bmatrix} M_d \\ M_q \end{bmatrix} \begin{bmatrix} C_\theta Z_{so11} - S_\theta Z_{so12} & C_\theta Z_{so12} + S_\theta Z_{so11} \end{bmatrix} \begin{bmatrix} C_\theta \\ S_\theta \end{bmatrix} \left(\tilde{m}_d I_{od} + \tilde{m}_q I_{oq}\right)$$
$$= Z_{so11} \begin{bmatrix} M_d \\ M_q \end{bmatrix} \begin{bmatrix} I_{od} & I_{oq} \end{bmatrix} \begin{bmatrix} \tilde{m}_d \\ \tilde{m}_q \end{bmatrix}. \quad \text{(A.10)}$$

Using (A.9) and (A.10) in (A.7), $\tilde{\mathbf{m}}\mathbf{V_{cf}} - \mathbf{M}\mathbf{Z_{so}}\tilde{\mathbf{m}}^T\mathbf{I_o}$ is evaluated as

$$\tilde{\mathbf{m}}\mathbf{V_{cf}} - \mathbf{M}\mathbf{Z_{so}}\tilde{\mathbf{m}}^T\mathbf{I_o} = \begin{bmatrix} \sqrt{\frac{3}{2}}\hat{V}_{cf}\cos\varphi_i - M_d Z_{so11} I_{od} & -M_d Z_{so11} I_{oq} \\ -M_q Z_{so11} I_{od} & \sqrt{\frac{3}{2}}\hat{V}_{cf}\cos\varphi_i - M_q Z_{so11} I_{oq} \end{bmatrix} \begin{bmatrix} \tilde{m}_d \\ \tilde{m}_q \end{bmatrix}$$
$$= \mathbf{G_a}(s) \begin{bmatrix} \tilde{m}_d \\ \tilde{m}_q \end{bmatrix}. \quad \text{(A.11)}$$

Denoting

$$a_V = \sqrt{\frac{3}{2}}\hat{V}_{cf}\cos\varphi_i, \quad \text{(A.12)}$$

$\mathbf{G_a}(s)$ in (A.11) can be expressed as,

$$\mathbf{G_a}(s) = \begin{bmatrix} a_V - M_d Z_{so11} I_{od} & -M_d Z_{so11} I_{oq} \\ -M_q Z_{so11} I_{od} & a_V - M_q Z_{so11} I_{oq} \end{bmatrix}. \quad \text{(A.13)}$$

A.3 Composition of $\mathbf{G_{vof}}$, $\mathbf{G_I^{-1}}(s)$, $\mathbf{G_a}(s)$

From (2.53) and (2.54) it is evident that all elements of transfer matrices which are directly transformed from their per phase scalar counter part have a common denominator polynomial. Using this knowledge and the same notations of (2.53), $\mathbf{Z_{so}}$, $\mathbf{Y_{mo}}$ and $\mathbf{G_{vof}}$ will be represented here as

$$\mathbf{Z_{so}} = \begin{bmatrix} Z_{so11} & Z_{so12} \\ -Z_{so12} & Z_{so11} \end{bmatrix} = \frac{1}{B_Z} \begin{bmatrix} N_{Z11} & N_{Z12} \\ -N_{Z12} & N_{Z11} \end{bmatrix}, \quad \text{(A.14)}$$

$$\mathbf{Y_{mo}} = \begin{bmatrix} Y_{mo11} & Y_{mo12} \\ -Y_{mo12} & Y_{mo11} \end{bmatrix} = \frac{1}{B_Y} \begin{bmatrix} N_{Y11} & N_{Y12} \\ -N_{Y12} & N_{Y11} \end{bmatrix}, \quad \text{(A.15)}$$

$$\mathbf{G_{vof}} = \begin{bmatrix} G_{vof11} & G_{vof12} \\ -G_{vof12} & G_{vof11} \end{bmatrix} = \frac{1}{B_{GVO}} \begin{bmatrix} N_{GVO11} & N_{GVO12} \\ -N_{GVO12} & N_{GVO11} \end{bmatrix}. \quad \text{(A.16)}$$

B_Z, B_Y and B_{GVO} are the common denominators of $\mathbf{Z_{so}}$, $\mathbf{Y_{mo}}$ and $\mathbf{G_{vof}}$ respectively which will be referred to as $C_d\,(\mathbf{Z_{so}})$, $C_d\,(\mathbf{Y_{mo}})$ and $C_d\,(\mathbf{G_{vof}})$.
Subsequently, $\mathbf{G_I^{-1}}(s)$ in (A.5) can be expressed as

$$\mathbf{G_I^{-1}} = \begin{bmatrix} 1+\frac{M_q N_{Z11}}{B_Z B_Y}\left(M_d N_{Y12}+M_q N_{Y11}\right) & -\frac{M_d N_{Z11}}{B_Z B_Y}\left(M_d N_{Y12}+M_q N_{Y11}\right) \\ -\frac{M_q N_{Z11}}{B_Z B_Y}\left(M_d N_{Y11}-M_q N_{Y12}\right) & 1+\frac{M_d N_{Z11}}{B_Z B_Y}\left(M_d N_{Y11}-M_q N_{Y12}\right) \end{bmatrix} \times \frac{1}{1+\frac{3}{4}m^2\frac{N_{Z11}N_{Y11}}{B_Z B_Y}}$$

$$= \begin{bmatrix} B_Z B_Y + M_q N_{Z11}\left(M_d N_{Y12}+M_q N_{Y11}\right) & -M_d N_{Z11}\left(M_d N_{Y12}+M_q N_{Y11}\right) \\ M_q N_{Z11}\left(M_d N_{Y11}-M_q N_{Y12}\right) & B_Z B_Y + M_d N_{Z11}\left(M_d N_{Y11}-M_q N_{Y12}\right) \end{bmatrix} \times \frac{1}{B_Z B_Y + \frac{3}{4}m^2 N_{Z11}N_{Y11}}. \quad (A.17)$$

Therefore all elements of $\mathbf{G_I^{-1}}$ have the common denominator

$$\begin{aligned} C_d\left(\mathbf{G_I^{-1}}\right) = B_Z B_Y + \frac{3}{4}m^2 N_{Z11}N_{Y11} = &\quad B_Z B_Y\left(1+\tfrac{3}{4}m^2 Z_{so11}Y_{mo11}\right) \\ &= C_d\left(\mathbf{Z_{so}}\right) C_d\left(\mathbf{Y_{mo}}\right)\left(1+\tfrac{3}{4}m^2 Z_{so11}Y_{mo11}\right). \end{aligned} \quad (A.18)$$

Determinant of the numerator matrix of $\mathbf{G_I^{-1}}$ can be evaluated as

$$\begin{aligned} det\left(\mathbf{N}\left(\mathbf{G_I^{-1}}(s)\right)\right) &= B_Z B_Y\left(B_Z B_Y + \left(M^2{}_d + M^2{}_q\right) N_{Z11}N_{Y11}\right) \\ &= \left(B_Z B_Y\right)^2\left(1+\tfrac{3}{4}m^2 Z_{so11}Y_{mo11}\right) \\ &= C_d\left(\mathbf{Z_{so}}\right) C_d\left(\mathbf{Y_{mo}}\right) C_d\left(\mathbf{G_I^{-1}}\right). \end{aligned} \quad (A.19)$$

From (A.13), $\mathbf{G_a}(s)$ can be represented as

$$\begin{aligned} \mathbf{G_a}(s) &= \begin{bmatrix} a_V - M_d Z_{so11} I_{od} & -M_d Z_{so11} I_{oq} \\ -M_q Z_{so11} I_{od} & a_V - M_q Z_{so11} I_{oq} \end{bmatrix} \\ &= \frac{1}{B_Z}\begin{bmatrix} B_Z a_V - M_d N_{Z11} I_{od} & -M_d N_{Z11} I_{oq} \\ -M_q N_{Z11} I_{od} & B_Z a_V - M_q N_{Z11} I_{oq} \end{bmatrix}. \end{aligned} \quad (A.20)$$

Therefore all elements of $\mathbf{G_a}(s)$ have the common denominator B_Z i.e. $C_d\left(\mathbf{Z_{so}}\right)$. Determinant of the numerator matrix is

$$\begin{aligned} det\left(\mathbf{N}\left(\mathbf{G_a}(s)\right)\right) &= B_Z^2 a_V^2 - B_Z a_V N_{Z11}\left(M_d I_{od} + M_q I_{oq}\right) \\ &= B_Z^2 a_V^2\left\{1 - Z_{so11}\frac{\left(M_d I_{od}+M_q I_{oq}\right)}{a_V}\right\} \\ &= C_d^2\left(\mathbf{Z_{so}}(s)\right) a_V^2\left\{1 - \frac{M_d I_{od}+M_q I_{oq}}{a_V} Z_{so11}\right\}. \end{aligned} \quad (A.21)$$

Appendix B
3 Phase Direct Matrix Converter Prototype

A 6 kVA prototype was built using the following discrete IGBT module with common emitter configuration.

Manufacturer:	SEMIKRON.
Part No.:	SK60GM123.
Maximum collector-emitter voltage:	1200 V.
Maximum continuous collector current:	40 A (80 °C), 60 A (25 °C).

The power stage layout was conceived on a single 6 layer PCB with each layer dedicated to one of the input or output phase. Six parallel layers of bus bar structure was laid out to minimize the stray inductance by reducing the current loop area to the maximum extent. It was designed to withstand 2.5 kV between consecutive layers and carry 15 A (rms) in each layer. Testing was carried out by applying 1.5 kV (rms) between each layer for one minute. Thereafter 15 A (rms) current was passed through all layers, with all of them connected in series, for one hour. No perceptible temperature rise was noticed. The circuit layout for clamp protection was also realized on the same PCB. This is shown in Fig. B.1. Figure B.2 shows the overall converter assembly with the IGBT gate driver cards, protection and level shifting (PSC) card, voltage sensor cards and fault indicator card. The gate drivers and input filter capacitors are mounted adjacent to the power stage PCB to minimize wiring inductance. The gate driver, voltage sensor, protection and signal level shifting circuits discussed in [57–59] has been modified and customized for the present work. The current sensor described in [60] has been adopted here.

A. Dasgupta and P. Sensarma, *Design and Control of Matrix Converters*,
Energy Systems in Electrical Engineering, DOI 10.1007/978-981-10-3831-0

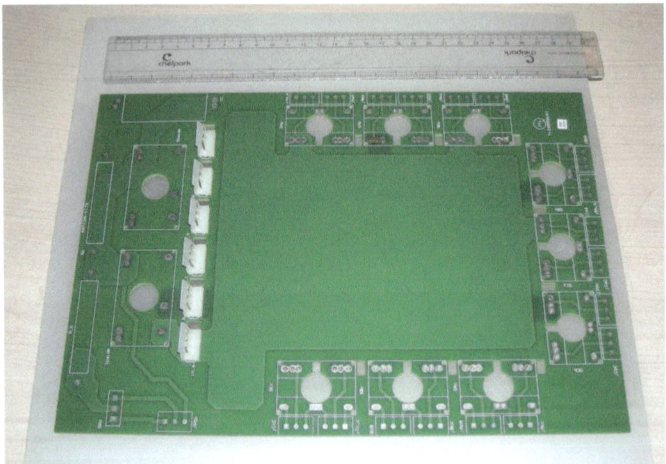

Fig. B.1 PCB housing the power stage and also the clamp protection circuit

Fig. B.2 Overall converter assembly with the gate driver, protection and sensor cards

Appendix C
Phase Locked Loop

A PLL reported in [51] has been followed in this work. Its working principle is briefly described in this Appendix.

A set of 3-phase voltages $\mathbf{v}^{(3)}$ composed of only fundamental frequency component is represented as

$$\mathbf{v}^{(3)} = \begin{bmatrix} v_a \\ v_b \\ v_c \end{bmatrix} = \hat{V} \begin{bmatrix} \cos(\omega t) \\ \cos(\omega t - 120^\circ) \\ \cos(\omega t + 120^\circ) \end{bmatrix} = \hat{V} \begin{bmatrix} \cos(\theta) \\ \cos(\theta - 120^\circ) \\ \cos(\theta + 120^\circ) \end{bmatrix}. \tag{C.1}$$

If these are transformed to a synchronously rotating dq domain using the transfer matrix

$$\mathbf{T_T} = \sqrt{\frac{2}{3}} \begin{bmatrix} \cos(\theta_e) & \cos(\theta_e - 120^\circ) & \cos(\theta_e + 120^\circ) \\ \sin(\theta_e) & \sin(\theta_e - 120^\circ) & \sin(\theta_e t + 120^\circ) \end{bmatrix}, \tag{C.2}$$

where θ_e is the estimated value of the phase angle θ, the resulting dq variables are

$$\mathbf{v} = \begin{bmatrix} v_d \\ v_q \end{bmatrix} = \sqrt{\frac{3}{2}} \hat{V} \begin{bmatrix} \cos(\theta_e - \theta) \\ \sin(\theta_e - \theta) \end{bmatrix} = \hat{E} \begin{bmatrix} \cos(\theta_e - \theta) \\ \sin(\theta_e - \theta) \end{bmatrix}. \tag{C.3}$$

Therefore, if the angular frequency ω can be correctly estimated, the transformed voltages are obtained as DC quantities. Moreover, if θ_e is accurately estimated then v_q becomes zero. If the estimation error is small, then (C.3) can be modified to

$$\mathbf{v} = \begin{bmatrix} v_d \\ v_q \end{bmatrix} \approx \hat{E} \begin{bmatrix} 1 \\ \theta_e - \theta \end{bmatrix}. \tag{C.4}$$

θ can now be estimated by using a linear feedback control scheme. Figure C.1 shows this estimation scheme as reported in [61].

A. Dasgupta and P. Sensarma, *Design and Control of Matrix Converters*, Energy Systems in Electrical Engineering, DOI 10.1007/978-981-10-3831-0

Fig. C.1 Estimation of θ

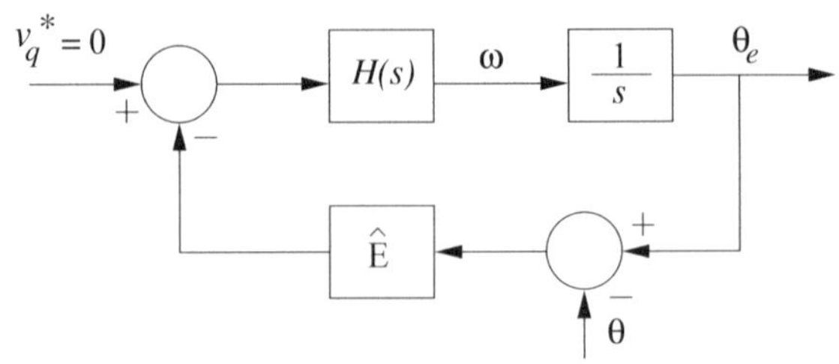

However, this approach cannot produce a fast yet accurate estimation of θ in presence of DC offset or negative sequence components or harmonics in $\mathbf{v}^{(3)}$. The presence of these components introduces a fundamental frequency component along with second harmonic and higher frequency components in v_q. To avoid θ_e from being affected by the unbalance in voltages or harmonic components, the closed loop bandwidth has to be low which in turn results in a slow locking time for the PLL.

To overcome this difficulty, an improved scheme has been proposed in [51] to extract the positive sequence component in the voltages before employing the basic estimation scheme depicted by Fig. C.1. Initially, $\mathbf{v}^{(3)}$ is transformed to dq domain by using a nominal frequency of 50 Hz for the rotating reference frame. Let the fundamental and harmonic frequency be denoted as f and f_h Hz respectively. Then, transformation to dq leads to introduction of $50 - f$ Hz, 50 Hz, $50 + f$ Hz and $50+f_h$ Hz components in v_d and v_q for the positive, DC offset, negative and harmonic frequency components respectively. These are then fed into a cascaded chain of filters as shown in Fig. C.2. Each link in the chain isdedicated to the removal of any one

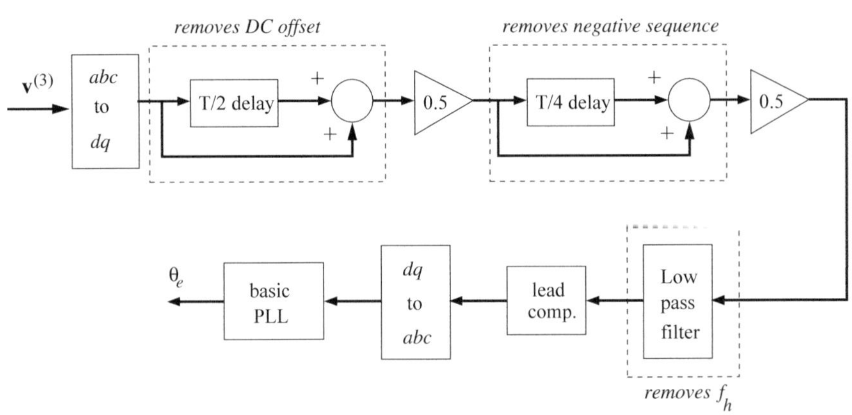

Fig. C.2 PLL proposed in [51]

of the frequency component other than $50 - f$ Hz corresponding to the positive sequence component. The period corresponding to the nominal frequency of 50 Hz is denoted by T. If the fundamental frequency differs from the nominal value, then the filtering process introduces a phase lag in the positive sequence components. Therefore, a lead compensator is used to correct this lag before transforming the *dq* variables back to *abc* domain which now contain only the positive sequence components. These voltages are now used for the basic PLL scheme to estimate θ_e.

References

1. D. Gopinath, Modeling, real-time simulation and design of matrix converters. Ph.D. Thesis, Indian Institute of Science, Bangalore, India. Sept. 2009
2. C. Klumpner, P. Nielsen, I. Boldea, F. Blaabjerg, New matrix converter motor (MCM) for industry applications. IEEE Trans. Ind. Electron. **49**(2), 325–335 (2002)
3. P. Wheeler, D. Grant, Optimised input filter design and low-loss switching techniques for a practical matrix converter. IEEE Proc. Elec. Pow. Appl. **144**(1), 53–59 (1997). Jan
4. C. Neft, C. Schauder, Theory and design of a 30-hp matrix converter. IEEE Trans. Ind. Appl. **28**(3), 546–551 (1992)
5. P. Nielsen, F. Blaabjerg, J.K. Pedersen, New protection issues of a matrix converter: design considerations for adjustable-speed drives. IEEE. Trans. Ind. Appl. **35**(5), 1150, 1161 (1999)
6. A. Trentin, P. Zanchetta, J. Clare, P. Wheeler, Automated optimal design of input filters for direct AC/AC matrix converters. IEEE Trans. Ind. Electron. **59**(7), 2811–2823 (2012). July
7. T. Kume, K. Yamada, T. Higuchi, E. Yamamoto, H. Hara, T. Sawa, M. Swamy, Integrated filters and their combined effects in matrix converter. IEEE Trans. Ind. Appl. **43**(2), 571–581 (2007)
8. D. Casadei, G. Serra, A. Tani, A. Trentin, L. Zarri, Theoretical and experimental investigation on the stability of matrix converters. IEEE Trans. Ind. Electron. **52**(5), 1409–1417 (2005). Oct
9. D. Casadei, J. Clare, L. Empringham, G. Serra, A. Tani, A. Trentin, P. Wheeler, L. Zarri, Large-signal model for the stability analysis of matrix converters. IEEE Trans. Ind. Electron. **54**(2), 939–950 (2007). Apr
10. J. Mahlein, M. Bruckmann, M. Braun, Passive protection strategy for a drive system with a matrix converter and an induction machine. IEEE Trans. Ind. Electron. **49**(2), 297–303 (2002)
11. N. Burany, Safe control of four-quadrant switches. *Conference Record of the Industry Applications Society Annual Meeting*, 1989, vol. 1, pp. 1190–1194, 1–5 Oct 1989
12. J. Oyama, T. Higuchi, E. Yamada, T. Koga, T. Lipo, New control strategy for matrix converter. *Power Electronics Specialists Conference, 1989. PESC '89 Record., 20th Annual IEEE*, vol. 1, pp. 360–367, 26–29 June 1989
13. P. Wheeler, J. Clare, L. Empringharn, M. Bland, M. Apap, Gate drive level intelligence and current sensing for matrix converter current commutation. IEEE Trans. Ind. Electron. **49**(2), 382–389 (2002)
14. K. Sun, D. Zhou, H. Lipei, K. Matsuse, K. Sasagawa, A novel commutation method of matrix converter fed induction motor drive using RB-IGBT. IEEE Trans. Ind. Appl. **43**(3), 777–786 (2007)
15. D. Gopinath, V. Ramanarayanan, *Implementation of bi-directional switch commutation scheme for matrix converters, in Proceedings National Power Electronics Conference-2007* (IISc Bangalore, India (CD), 2007)

A. Dasgupta and P. Sensarma, *Design and Control of Matrix Converters*,
Energy Systems in Electrical Engineering, DOI 10.1007/978-981-10-3831-0

16. J. Mahlein, J. Igney, J. Weigold, M. Braun, O. Simon, Matrix converter commutation strategies with and without explicit input voltage sign measurement. IEEE Trans. Ind. Electron. **49**(2), 407–414 (2002). Apr
17. H. She, H. Lin, B. He, X. Wang, L. Yue, X. An, Implementation of voltage-based commutation in space vector modulated matrix converter. IEEE Trans. Ind. Electron. **59**(1), 154–166 (2012). Jan
18. R. Cardenas, R. Pena, P. Wheeler, J. Clare, G. Asher, Control of the reactive power supplied by a WECS based on an induction generator fed by a matrix converter. IEEE Trans. Ind. Electron. **56**(2), 429–438 (2009). Feb
19. R. Vargas, U. Ammann, B. Hudoffsky, J. Rodriguez, P. Wheeler, Predictive torque control of an induction machine fed by a matrix converter with reactive input power control. IEEE Trans. Pow. Electron. **25**(6), 1426–1438 (2010). June
20. L. Wei, T. Lipo, A novel matrix converter topology with simple commutation. *Industry Applications Conference, 2001. Thirty-Sixth IAS Annual Meeting. Conference Record of the 2001 IEEE*, vol. 3, pp. 1749–1754, 30 Sep.–4 Oct. 2001
21. J. Kolar, T. Friedli, J. Rodriguez, P. Wheeler, Review of three-phase PWM AC-AC converter topologies. IEEE Trans. Ind. Electron. **58**(11), 4988–5006 (2011). Nov
22. A. Alesina, M. Venturini, Solid-state power conversion: a Fourier analysis approach to generalized transformer synthesis. IEEE Trans. Circuits. Syst. **28**(4), 319–330 (1981). Apr
23. A. Alesina, M. Venturini, Analysis and design of optimum-amplitude nine-switch direct AC-AC converters. IEEE Trans. Pow. Electron. **4**(1), 101–112 (1989). Jan
24. P. Ziogas, S. Khan, M. Rashid, Some improved forced commutated cycloconverter structures. IEEE Trans. Ind. Appl. **IA-21**(5), 1242–1253 (1985)
25. L. Huber, D. Borojevic, Space vector modulated three phase to three phase matrix converter with input power factor correction. IEEE Trans. Ind. Appl. **31**(6), 1234–1246 (1995)
26. H. Cha, P. Enjeti, An approach to reduce common-mode voltage in matrix converter. IEEE Trans. Ind. Appl. **39**(4), 1151–1159 (2003)
27. D. Casadei, G. Serra, A. Tani, L. Zarri, Optimal use of zero vectors for minimizing the output current distortion in matrix converters. IEEE Trans. Ind. Electron. **56**(2), 326–336 (2009). Feb
28. L. Helle, K. Larsen, A. Jorgensen, S. Munk-Nielsen, F. Blaabjerg, Evaluation of modulation schemes for three-phase to three-phase matrix converters. IEEE Trans. Ind. Electron. **51**(1), 158–171 (2004). Feb
29. K. Mohapatra, P. Jose, A. Drolia, G. Aggarwal, S. Thuta, N. Mohan, A novel carrier-based PWM scheme for matrix converters that is easy to implement. *Power Electronics Specialists Conference, 2005. PESC '05. IEEE 36th*, pp. 2410–2414, June 2005
30. Y. Yoon, S. Sul, Carrier-based modulation technique for matrix converter. IEEE Trans. Pow. Electron. **21**(6), 1691–1703 (2006). Nov
31. S. Muller, U. Ammann, S. Rees, New time-discrete modulation scheme for matrix converters. IEEE Trans. Ind. Electron. **52**(6), 1607–1615 (2005). Dec
32. J. Rodriguez, M. Rivera, J. Kolar, P. Wheeler, A review of control and modulation methods for matrix converters. IEEE Trans. Ind. Electron. **59**(1), 58–70 (2012). Jan
33. H. Nikkhajoei, A. Tabesh, M. Iravani, Dynamic model of a matrix converter for controller design and system studies. IEEE Trans. Pow. Del. **21**(2), 744–754 (2006). Apr
34. B. Wang, G. Venkataramanan, Dynamic voltage restorer utilizing a matrix converter and flywheel energy storage. IEEE Trans. Ind. Appl. **45**(1), 222–231 (2009)
35. D. Casadei, G. Serra, A. Tani, and L. Zarri, Stability analysis of electrical drives fed by matrix converters. *Proceedings IEEE ISIE*, pp. 1108–1113, July 8–11, 2002
36. F. Liu, C. Klumpner, F. Blaabjerg, Stability analysis and experimental evaluation of a matrix converter drive system. *Industrial Electronics Society, 2003. IECON '03. The 29th Annual Conference of the IEEE*, vol. 3, pp. 2059–2065, 2–6 Nov. 2003
37. F. Liu, C. Klumpner, F. Blaabjerg, A robust method to improve stability in matrix converters. Proc. IEEE PESC'04. **6**, 3560–3566 (2004)
38. R. Cardenas, R. Pela, G. Tobar, J. Clare, P. Wheeler, G. Asher, Stability analysis of a wind energy conversion system based on a doubly fed induction generator fed by a matrix converter. IEEE Trans. Ind. Electron. **56**(10), 4194–4206 (2009). Oct

39. P. Zanchetta, P. Wheeler, L. Empringham, J. Clare, Design control and implementation of a three-phase utility power supply based on the matrix converter. IET Pow. Electron. **2**(2), 156–162 (2009)
40. J. Monteiro, J. Silva, S. Pinto, J. Palma, Matrix converter-based unified power-flow controllers: advanced direct power control method. IEEE Trans. Pow. Del. **26**(1), 420–430 (2011)
41. K. Padiyar, A. Kulkarni, Design of reactive current and voltage controller of static condenser. Electric. Pow. Energy. Syst. **19**(6) (1997)
42. A. MacFarlane, N. Karcanias, Poles and zeros of linear multivariable systems: a survey of the algebraic, geometric and complex-variable theory. Int. J. Control. **24**(1), 33–76 (1976)
43. T. Kailath, *Linear Systems* (Prentice-Hall Inc., 1980)
44. S. Skogestad, I. Postlethwaite, *Multivariable Feedback Control*, 2nd edn. (Wiley, 2005)
45. P. Albertos, A. Sala, *Multivariable Control Systems: An Engineering Approach* (Springer-Verlag, London Ltd., 2004)
46. R. Erickson, D. Maksimovic, *Fundamentals of Power Electronics*, 2nd edn. (Springer, 2001)
47. N. Mohan, T. Underland, W. Robbins, *Power Electronics: Converters, Applications, and Design*, 2nd edn. (Wiley, NY, 2001)
48. M. Bollen, *Understanding Power Quality Problems: Voltage Sags and Interruptions* (Wiley-IEEE press, 1999)
49. R. Dugan, M. McGranaghan, S. Santoso, H. Wayne Beaty, *Electrical Power Systems Quality*, 2nd edn. (McGraw-Hill, 2003)
50. J. Nielsen, F. Blaabjerg, N. Mohan, Control strategies for dynamic voltage restorer compensating voltage sags with phase jump. Proc. IEEE-APEC 1267–1273 (2001)
51. R. Sinha, P. Sensarma, Improved PLL under distorted utility conditions. Proc. IEEE-ICIT 1849–1854 (2006)
52. IEEE Recommended Practice for Evaluating Electric Power System Compatibility With Electronic Process Equipment, IEEE Std. 1346–1998, 1998
53. W. Rohouma, L. Empringham, P. Zanchetta, P. Wheeler, A four legs matrix converter based ground power unit with selective harmonic control. *Energy Conversion Congress and Exposition (ECCE), 2011 IEEE*, pp. 799–805, 17–22 Sept. 2011
54. P. Wheeler, P. Zanchetta, J. Clare, L. Empringham, M. Bland, D. Katsis, A utility power supply based on a four-output leg matrix converter. IEEE Trans. Ind. Appl. **44**(1), 174–186 (2008)
55. T. Friedli, J.W. Kolar, J. Rodriguez, P.W. Wheeler, Comparative evaluation of three-phase AC-AC matrix converter and voltage DC-link back-to-back converter systems. IEEE Trans. Ind. Electron. **59**(12), 4487–4510 (2012). Dec
56. L. Empringham, J.W. Kolar, J. Rodriguez, P.W. Wheeler, J.C. Clare, Technological issues and industrial application of matrix converters: a review. IEEE Trans. Ind. Electron. **60**(10), 4260–4271 (2013). Oct
57. B. Sen, *Design, simulation and realization of a wideband shunt hybrid active filter* (M. Tech, Thesis, Indian Institute of Technology, Kanpur, India, 2006)
58. N.V.N.S. Srinivas, *Design, development and implementation of minimum VA rated UPQC* (M. Tech, Thesis, Indian Institute of Technology, Kanpur, India, 2006)
59. R. Bachana, *Improvement of current dynamics during controller saturation in a D-STATCOM* (M. Tech, Thesis, Indian Institute of Technology, Kanpur, India, 2006)
60. R. Sinha, *Double vector control of dynamic voltage restorer using improved phase locked loop* (M. Tech, Thesis, Indian Institute of Technology, Kanpur, India, 2006)
61. V. Kaura, V. Blasko, Operation of a phase locked loop system under distorted utility conditions. IEEE Trans. Ind. Appl. **33**(1), 58–63 (1997)
62. P. Wheeler, J. Rodriguez, J. Clare, L. Empringham, A. Weinstein, Matrix converters: a technology review. IEEE Trans. Ind. Electron. **49**(2), 276–288 (2002)
63. S. Barakati, M. Kazerani, J. Aplevich, An overall dynamic model for a matrix converter. Proc. IEEE ISIE 3–18 (2008)
64. N. Sokal, System oscillations caused by negative input resistance at the power input port of a switching mode regulator, amplifier, dc/dc converter, or dc/ac inverter. *Conference Record of IEEE PESC*, 1973, pp. 138–140

65. R. Middlebrook, Input filter considerations in design and application of switching regulators *Conference Record IEEE Industrial Application Society Annual Meeting*, 1976, pp. 366–382
66. A. Dasgupta, P. Sensarma, Low-frequency dynamic modelling and control of matrix converter for power system applications. IET Pow. Electron. **5**(3), 304–314 (2012)
67. J. Nielsen, M. Newman, H. Nielsen, F. Blaabjerg, Control and testing of a dynamic voltage restorer (DVR) at medium voltage level. IEEE Trans. Pow. Electron. **19**(3), 806–813 (2004)

MIX
Papier aus verantwortungsvollen Quellen
Paper from responsible sources
FSC® C105338

If you have any concerns about our products,
you can contact us on
ProductSafety@springernature.com

In case Publisher is established outside the EU,
the EU authorized representative is:
Springer Nature Customer Service Center GmbH
Europaplatz 3, 69115 Heidelberg, Germany

Printed by Libri Plureos GmbH
in Hamburg, Germany